AF743455

FINTECH AND BLOCKCHAIN TECHNOLOGY IN FINANCIAL MARKETS

Electronic Commerce Management for Business

FINTECH AND BLOCKCHAIN TECHNOLOGY IN FINANCIAL MARKETS

Neha Puri, PhD
Anjali Munde, PhD
Emilia Alaverdov, PhD
Editors

AAP | APPLE ACADEMIC PRESS

First edition published 2027

Apple Academic Press Inc.
1265 Goldenrod Circle, NE,
Palm Bay, FL 32905 USA

760 Laurentian Drive, Unit 19,
Burlington, ON L7N 0A4, Canada

CRC Press
2385 NW Executive Center Drive,
Suite 320, Boca Raton FL 33431

4 Park Square, Milton Park,
Abingdon, Oxon, OX14 4RN, UK

© 2027 by Apple Academic Press, Inc.

Apple Academic Press exclusively co-publishes with CRC Press, an imprint of Taylor & Francis Group, LLC

Reasonable efforts have been made to publish reliable data and information, but the authors, editors, and publisher cannot assume responsibility for the validity of all materials or the consequences of their use. The authors, editors, and publishers have attempted to trace the copyright holders of all material reproduced in this publication and apologize to copyright holders if permission to publish in this form has not been obtained. If any copyright material has not been acknowledged, please write and let us know so we may rectify in any future reprint.

Except as permitted under U.S. Copyright Law, no part of this book may be reprinted, reproduced, transmitted, or utilized in any form by any electronic, mechanical, or other means, now known or hereafter invented, including photocopying, microfilming, and recording, or in any information storage or retrieval system, without written permission from the publishers.

For permission to photocopy or use material electronically from this work, access www.copyright.com or contact the Copyright Clearance Center, Inc. (CCC), 222 Rosewood Drive, Danvers, MA 01923, 978-750-8400. For works that are not available on CCC please contact mpkbookspermissions@tandf.co.uk

Trademark notice: Product or corporate names may be trademarks or registered trademarks and are used only for identification and explanation without intent to infringe.

For Product Safety Concerns and Information please contact our EU representative GPSR@taylorandfrancis.com Taylor & Francis Verlag GmbH, Kaufingerstraße 24, 80331 München, Germany

Library and Archives Canada Cataloguing in Publication

CIP data on file with Canada Library and Archives

Library of Congress Cataloging-in-Publication Data

CIP data on file with US Library of Congress

ISBN: 978-1-77964-128-1 (hbk)
ISBN: 978-1-77964-129-8 (ebk)
DOI: 10.1201/9781779641298

About the Book Series Electronic Commerce Management for Business

Vikas Garg, PhD
Associate Professor, Symbiosis Institute of Business Management, Noida, Uttar Pradesh, India

Richa Goel, PhD
Symbiosis Centre for Management Studies, Noida, India

Electronic trading, typically called e-commerce, involves procurement and distribution of electronic systems and other digital networks of goods or services. E-commerce stands for electronic commerce and concerns the electronic medium trade in goods and services. As the Internet grew, the volume of online trade has increased significantly. B2B, B2C, C2C, and related resources help to build online technology for future customer needs and markets. E-commerce has revolutionized industry, disrupted trade with the Internet, and created an e-commerce market for customers and enterprises through the electronic communications system.

The advances in internet and web-based technology have steadily eroded boundaries between traditional economies and the new online economy, including the amount of corporate money. The growth in e-commerce is enormous. Another explanation for this is the low cost of the PC and the rising use of the Internet. The corporate community is rapidly conscious of the possibilities provided by e-commerce.

This series mainly discusses the role of e-commerce in today's business. It further highlights some critical issues in e-commerce, suggestions, and future strategies for e-commerce in years to come.

Topics of Books in the Series:

The primary endeavor of this series is to introduce and explore contemporary research developments in a variety of rapidly growing research areas. The volumes deal with the following topics but not limited to:

- E-Business and Mobile Commerce

- Big Data Analytics
- Business and Data Analytics
- Implications of Technological Disruption
- Technology and the War for Talent
- Impact of New Technologies on Business
- Artificial Intelligence and Its Impact on Business
- Managing Global Competitiveness Using Technology
- Impact of New Technology on Global Labor Markets
- Blockchain in International Business
- IoT and Smart Cities
- IoT Strategies and Business models
- Industrial Internet of Things (IoT)
- AR (Augmented Reality) in Product Design and Manufacturing
- AR (Augmented Reality) and Training
- Use of AR (Augmented Reality) in Sales and Marketing
- Smart Manufacturing
- Cryptocurrency
- Strategies of Emerging Countries in the Era of New Technology
- The Roadmap to Thriving in the New Skills Economy
- Future of Mobile Apps
- Social Media – A Powerful Tool
- Big Data Powers Organizations
- Integration of Technology in HR Practices
- Skillset for Industry 4.0
- Managing Virtual Teams and Environments for Innovation
- Innovation in Business
- Effects of Shifts in Globalization on Emerging Economies
- Smart Global Supply Chain Management
- Corporate Governance in the New Global Technology-Driven Economy
- Business Intelligence Tools & Applications
- Cyber Defense & Information Technology
- Cloud Computing
- Data & Intelligent System Modelling
- Digital Marketing & Internet Marketing
- Smart Connected Products
- Impact of Technology on Financial Decisions
- Finance and Risk Management for Using New Technology

- Digital Transformation and Financial Services
- Banking services in the digital era
- Virtual Reality
- Robotics and Its Use in Business
- Industry 4.0 and Labor Issues

The series serves as a valuable resource for researchers, academicians, and industry professionals and undergraduate, postgraduate, doctoral students in electronic commerce, Internet marketing, business innovation, digitization, technology in product design, management, and more.

Features of the volumes under this series explore recent trends, model extensions, developments, solutions to real-time problems, and case studies.

BOOKS IN THE SERIES

Applications of Artificial Intelligence in Business and Finance 5.0
Editors: Richa Goel, PhD, Vikas Garg, PhD, and Michela Floris, PhD

Blockchain Technology: Unlocking the Value of 3Ts in Business World
Editors: Richa Chauhan, PhD, and Vidhi Kaul, PhD

Entrepreneurship and Sustainable Business Development
Editors: Ruchi Khandelwal, PhD, Deepa Kapoor, PhD, Nidhi Gupta, PhD, and Neha Gupta, PhD

The Role of Smart Technology and Industry 4.0 in Shaping Smart Cities
Editors: Vinay Kandpal, PhD, Manoj Kumar Nallapaneni, PhD, and Deep Chandra, PhD

Fintech and Blockchain Technology in Financial Markets
Editors: Neha Puri, PhD, Anjali Munde, PhD, and Emilia Alayerdov, PhD

E-Financial Strategies for Sustainable Development: A Handbook for Promoting Financial Inclusion and Ending Poverty (SDG 1)
Editors: Ernesto D. R. Santibanez Gonzalez, PhD, Rajeev Sijaria, PhD, and Vikas Garg, PhD

Green Finance and Sustainable Development Goals
Editors: Vikas Garg, PhD, Krishnendu Ghosh, PhD, and Ronita Ram, PhD

Industry 5.0: Innovation and Adoption for Sustainability
Editors: Tilottama Singh, PhD, Pradeep Suri , PhD, and Zulkarin Jahangir, PhD

Global Sustainable Development: Integration of Digital and Human-Centric Pathways
Editors: Kavita Indapurkar, PhD, Neeru Sidana, PhD, and
Iqbal Thonse Hawaldar, PhD

About the Editors

Neha Puri, PhD
Associate Professor, Amity College of Commerce & Finance (ACCF), Amity University, Noida, Uttar Pradesh, India

Neha Puri, PhD, is an accomplished Associate Professor at Amity College of Commerce & Finance (ACCF), Amity University, Noida, Uttar Pradesh, India, with over 15 years of teaching experience at esteemed institutions. She is a dedicated academician, serving as the chief course coordinator for various financial accounting courses and as a member of the boards of studies. Her expertise spans multiple areas in finance and accounting, including financial statement analysis, financial performance, and accounting standards. An avid researcher, she has numerous publications in international journals on topics such as IFRS (International Financial Reporting Standards), financial statement analysis, oil prices, and exchange rate volatility. She is currently supervising PhD scholars and is recognized as a proficient reviewer for reputable journals. She has been invited to serve as a session chair at various international conferences and as a resource person for faculty development programs on bibliometric analysis as well as training programs for bankers. Her accolades include best research paper awards, and she has registered two copyrights.

Anjali Munde, PhD
University of Southampton, Malaysia Campus, Malaysia

Anjali Munde, PhD, holds a bachelor's degree in Mathematics (Honors) and a master's degree in Mathematics from Hans Raj College, University of Delhi, India, earned in 2004 and 2006, respectively. She was awarded the prestigious Gold Medal from the same university. In 2016, she obtained her PhD in Mathematics. Presently, she is with the University of Southampton, Malaysia Campus. Before her current assignment, she taught at renowned institutions such as the School of Inspired Leadership, Gurugram; Amity University, Noida; and IIMT, G.G.S.I.P. University,

Delhi. She obtained a Certificate in Data Science and Business Analytics from Great Learning, Great Lakes Institute of Management, in 2021. Dr. Munde has published several research papers in well-known journals indexed in ABDC, Scopus (Elsevier), IEEE, Web of Science, and Springer, and has presented technical papers at various national and international conferences. She has also published book chapters indexed in ScienceDirect and Web of Science. With over 17 years of teaching experience, her research interests include fuzzy information measures, decision making, coding theory, business analytics, machine learning, and data science. She serves as a reviewer and editorial board member for several national and international journals, including the *International Journal of General Systems* (Taylor & Francis), *International Journal of Information Systems in the Service Sector* (IGI Global), *Journal of Mathematics and Statistics* (Elsevier), and *International Journal of Mathematics and Mathematical Sciences* (Wiley). She is a life member of the International Association of Engineers (IAENG), India, and has served as an editorial board member for senior secondary courses at the National Institute of Open Schooling (NIOS), under the Ministry of Human Resource Development, Government of India. Dr. Munde has conducted numerous faculty development programs (FDPs) and workshops in data analytics, machine learning, and Python, organized by ICMAI Registered Valuers Organisation (recognized RVO under the Insolvency and Bankruptcy Board of India) promoted by The Institute of Cost Accountants of India (statutory body under an Act of Parliament), G.G.S.I.P. University, Delhi; G.H. Raisoni University; and Sri Krishna College of Technology, Coimbatore. She has also served as a session chair, technical program committee member, advisory committee member, and reviewer at various international conferences organized by the National Institute of Technology, Indian Institute of Information Technology, Christ University, ICFAI Business School, and others.

Dr. Munde has supervised four PhD research scholars. Her key teaching areas include quantitative techniques, statistics for management, operations research, business mathematics, and business analytics, using MS Excel, Python, and Tableau.

Emilia Alaverdov, PhD
Professor and Head, International Relations Bachelor's program, Faculty of Law and International Relations, Georgian Technical University, Tbilisi, Georgia

Emilia Alaverdov, PhD in Political Science, is a Professor and Head of the International Relations Bachelor's program at the Faculty of Law and International Relations, Georgian Technical University, Tbilisi, Georgia. She earned her bachelor's degree in Languages and Literature from Ilia Chavchavadze State University, Tbilisi, Georgia, with a qualification as a philologist of English Language and Literature and Spanish Language and Literature. She holds a master's degree in international relations and politics from the Diplomatic Academy of Georgia, where she began her academic career. Dr. Alaverdov is a member of the editorial boards of 12 international journals, editor of nine books, author of about 31 scientific papers, and co-author of seven books. She has participated in eight Erasmus+ projects and more than 50 international scientific conferences, the majority of them abroad, and has organized various international scientific events. Her research interests include religion, migration, religious tourism, and European and Russian studies. She is fluent in English, Russian, Georgian, Azerbaijani, Armenian, and Greek.

Contents

Contributors

Adarsh Arora
Amity College of Commerce & Finance (ACCF), Amity University, Noida, UP, India

Rashi Baliyan
Noida International University, Greater Noida, Uttar Pradesh, India

Sandeep Bansal
Indira Gandhi National College, Ladwa, Kurukshetra, Haryana, India

Divya Choudhary
GNIOT MBA Institute, Greater Noida, Uttar Pradesh, India

Vikas Garg
Symbiosis Institute of Business Management, Noida, Uttar Pradesh, India

Chabi Gupta
Amity College of Commerce & Finance, Amity University, Noida, UP, India

Suhasini Jain
Amity College of Commerce & Finance (ACCF), Amity University, Noida, Uttar Pradesh, India

Jyotsna
Jagan Institute of Management Studies, New Delhi, Delhi, India

Simrat Kaur
Jagannath International Management School, New Delhi, India

Shivi Mittal
Mangalmay Institute of Management & Technology, Greater Noida, Uttar Pradesh, India

Anjali Munde
Southampton Malaysia Business School, University of Southampton Malaysia, Iskandar Puteri, Johor, Malaysia

Sunil Munde
Engineering Management - Topside Engineering Coordination, Kuala Lumpur, Malaysia

Mohit Narang
Amity College of Commerce & Finance (ACCF), Amity University, Noida, Uttar Pradesh, India

Fincy Pallissery
School of Law, CHRIST University Bangalore

Ankur Pandey
Aadhar Housing Finance Ltd. Mumbai, Maharashtra, India

Kajal Panwar
Amity College of Commerce & Finance (ACCF), Amity University, Noida, Uttar Pradesh, India

Kriti Parashar
School of Law, CHRIST University Bangalore, India

Chanchal Phore
Maharaja Surajmal Institute, New Delhi, Delhi, India

Priyanka
Maharaja Surajmal Institute, New Delhi, Delhi, India

Neha Puri
Amity College of Commerce & Finance (ACCF), Amity University, Noida, Uttar Pradesh, India

Pallavi Rajain
Maharaja Surajmal Institute, New Delhi, Delhi, India

Subhadra Rajpoot
Amity University, Greater Noida, Uttar Pradesh, India

Namita Rajput
Sri Aurobindo College, University of Delhi, Delhi, India

Anshika Rajvanshi
IMS, Ghaziabad, Uttar Pradesh, India

Richa Srivastava
Noida International University, Greater Noida, Uttar Pradesh, India

Deepak Tandon
Senior Professor (Finance and Accounting), Institute of Management Studies (IMS), Ghaziabad, Uttar Pradesh, India

RajKumar Teotia
H. R. Institute of Science and Technology, Ghaziabad, Uttar Pradesh, India

Nisha Tokas
Maharaja Surajmal Institute, New Delhi, Delhi, India

Ayushi Tomar
Maharaja Surajmal Institute, New Delhi, Delhi, India

Om Prakash Yadav
School of Business Management, Noida International University, Greater Noida, Uttar Pradesh, India

Foreword

Technology and established financial institutions working together have ushered in a new era of innovation and opportunity in the fast-paced world of contemporary finance. FinTech and blockchain technology are leading this transformation, changing how we perceive value, invest, and conduct transactions. Businesses, individuals, and economies must understand and take advantage of the opportunities presented by these advancements—an imperative that becomes increasingly clear as we navigate this complex landscape.

Modern technology and traditional financial principles are converging to create a seismic shift in financial markets. FinTech has transformed the way we invest, transact, and manage our finances through its suite of digital tools and innovative business models. Simultaneously, blockchain technology, with its immutable ledger and decentralized architecture, promises to revolutionize the very fabric of our financial systems.

In this carefully curated exploration, we delve into the trends and applications shaping the landscape of FinTech and blockchain in financial markets. From the rise of mobile payments and robo-advisors to the advent of decentralized finance (DeFi) and asset tokenization, each chapter reveals a new facet of this transformative journey.

Through insightful analysis and real-world case studies, we uncover the profound impact of these technologies on traditional banking, investment management, insurance, and beyond. We see how FinTech is democratizing access to financial services, empowering individuals and businesses to take control of their financial futures. Likewise, we recognize the disruptive potential of blockchain technology, as it lays the foundation for a more transparent, secure, and inclusive financial ecosystem.

Yet, alongside the promise of innovation, we must address the challenges and complexities that accompany such rapid change. Regulatory hurdles, cybersecurity threats, interoperability issues, and scalability concerns loom large, demanding thoughtful solutions and collaborative efforts from all stakeholders.

May this book serve as a guiding light on your journey into the world of FinTech and blockchain in financial markets. May it inspire you to embrace innovation, navigate uncertainty, and seize the opportunities that lie ahead. I hope it will be a source of inspiration and knowledge as you navigate the ever-evolving financial landscape with confidence and foresight. Congratulations to the editors and writers for compiling such an impactful collection and for recognizing the promise of this topic early on.

—Sharuti Kapoor
IT Program Manager
City National Bank, California, USA

Acknowledgments

First and foremost, we are thankful to Divinity for giving us the ability to write and for leading us to pursue our hobbies and aspirations. This endeavor could not have been accomplished without confidence.

Sincere gratitude is also owed to our families, whose steadfast support helped us through the difficult process of finishing this book.

The writers and reviewers who devoted their time and knowledge to this project are highly appreciated, and the editors would like to thank everyone who was involved. This book would never have happened without their tremendous efforts.

To begin with, we thank all of the authors for their important contributions. We especially thank the chapter authors for their hard work and expertise.

Second, we would like to express our gratitude to the reviewers for their insightful comments that improved the chapters' quality, consistency, and layout. We appreciate that they served as both authors and referees.

We owe a debt of gratitude to everyone who offered constructive criticism that made us rethink ideas and consider different angles of our work. We also express our gratitude to everyone who has helped us along the way as we have studied and applied the art and science of networking.

—***Editors***

Preface

In recent years, financial technology (FinTech) has been front-page news. In its broadest sense, FinTech refers to the use of technological innovations in financial services and processes. Global FinTech investments totaled $210 billion in 2021, according to recent annual figures. In this developing landscape, start-ups, major technology corporations, and banks have emerged as significant stakeholders. Policymakers and regulators view FinTech as a means of making the financial system more efficient, effective, and resilient. Blockchain can help address the challenges of preventing financial exclusion. FinTech is rapidly changing the way financial services are delivered. Artificial intelligence (AI), blockchain, cloud computing, data analytics, and the Internet of Things (IoT) are among the enabling technologies for such innovation. FinTech operations take many forms and cover a wide range of financial services. Technology can improve these services by lowering costs, enhancing decision-making and execution, and expanding access. Significant decreases in the cost of processing, storage, and devices, for example, make it easier to analyze large volumes of data at low cost, thereby improving the accessibility of financial services. Since 2015, the potential of blockchain as a new means to generate, exchange, and monitor ownership of financial assets has captured the industry's attention. Smart contracts—computerized protocols containing terms dependent on decentralized consensus, tamperproof, and self-enforcing through automatic execution—have also emerged through blockchain technology. This book provides an outline of the FinTech environment as well as the topics that may be explored due to blockchain's new digital advancements. FinTech players such as Visa, Mastercard, and major financial institutions, for their part, are already conducting research and making their findings publicly available.

Moreover, crowdfunding research is the only branch of FinTech that shows significant interconnections. At both local and global levels, other FinTech research streams remain highly fragmented, with few interrelations. Despite its relative youth, blockchain has emerged as a sector worthy of further investigation within this fragmented FinTech research landscape. Blockchain is believed to have the potential to revolutionize investing and disrupt the global financial system. Currently, there are about 49 rapidly growing start-ups working on blockchain technology. FinTech refers to digital advancements and technologically enabled business model innovations in the financial industry, and it is likely that soon no industry will remain untouched by blockchain. This technology is already applied in diverse fields, including law, political science, policy, economics, the military, self-driving vehicles, education, and space exploration. Financial technology, or FinTech, represents the merging of technology and finance. Today's FinTech is driven by two major trends. The first is the rapid pace of change, fueled by big data, machine learning, technological commoditization, and artificial intelligence (AI). The second is the growing participation and investment of non-financial companies in the financial services industry. FinTech is an important sector in the development of Industry 4.0 because it requires the integration of many technologies, such as artificial intelligence and data science, and it serves as a platform for innovation. Emerging sectors based on machine learning techniques, such as learning analytics, offer practical opportunities to develop blockchain applications that can help instructors and students adopt diverse classroom and teaching methods. As many educational institutions seek to attract more international students, broaden access, increase population diversity, and scale their offerings, such AI systems may become essential components of educational infrastructure and critical tools for achieving future goals in the education sector. The future use of blockchain technology in numerous industries will make systems more interactive, customized, and engaging. Virtual reality, haptics, and companion robots are also expected to enter the home as hardware and sensor technologies advance.

CHAPTER 1

Emerging Trends in the Financial Market Using Blockchain Technology: A Bibliometric Review

SIMRAT KAUR[1] and ADARSH ARORA[2*]

[1]*Jagannath International Management School, New Delhi, India*

[2]*Amity College of Commerce & Finance (ACCF), Amity University, Noida, UP, India*

*Corresponding author

ABSTRACT

The health of a country's financial market affects its economic situation. A financial market is a sophisticated system that involves the interactions of many different parties, including buyers, sellers, financial advisors, and others.

The chapter describes the characteristics of blockchain technology in the context of the current financial market. In order to advance the uniformity and inventive development of blockchain in financial markets, this chapter elaborates on the innovation path of blockchain in the financial markets.

This review's main objective is to discover new research directions in the use of blockchain technology in the financial markets. This study seeks to review a few of the innovative research on blockchain technology in the financial market and direct future research in this area. By thoroughly employing bibliometric analysis to look at the connections between different domains, our study advances the body of knowledge. A conceptual perspective on this certain subject area has been identified through this review.

The main objective is to conduct thematic analysis, keyword co-occurrence analysis, and bibliographic coupling analysis on significant

FinTech and Blockchain Technology in Financial Markets. Neha Puri, Anjali Munde, Emilia Alaverdov (Eds.)
© 2027 Apple Academic Press, Inc. Co-published with CRC Press (Taylor & Francis)

DOI: 10.1201/9781779641298-1

published research articles. Scopus is being taken into consideration as the search engine for data gathering. The VOS Viewer and Biblioshiny tools are used in this review.

1.1 INTRODUCTION

The financial sector is essential for advancing the growth of the country's economy, and the risk it assumes also represents the biggest, unexpected risk in the financial market. To ensure that financial markets operate as intended, the risk present in the current financial market should be thoroughly investigated using blockchain technology (Liu and Yu, 2022).

The most significant technological advancement of the 21st century is blockchain technology. The study of blockchain technology's potential applications in the financial sector is progressing along with network technology, and the benefits and features of this technology are critical to the modernization and optimization of the established financial system (Zhang, 2022). The financial market has given close attention to the rapid development of blockchain technology, which is very important for the financial market's overall risk aversion. A blockchain is a distributed record that's made up of blocks with transaction information that is linked chronologically to form a chain (Sidiq et al., 2022).

1.2 OBJECTIVES

- To profile the literature to discover how financial markets using blockchain technology journals, papers, and authors are distributed.
- To ascertain how financial market documents using blockchain technology are bibliographically coupled.
- To conduct document thematic analysis to examine the novel paths that have been adopted by the financial markets using blockchain technology.
- To perform a co-occurrence analysis of keywords in financial markets papers using blockchain technology.

1.3 LITERATURE REVIEW

The financial world is growing increasingly complex in the digital age due to the advancement and application of information technology, and traditional financial industry products and service models are progressively becoming inadequate to fulfill people's everyday needs. In order to address this issue, it is critical to progressively improve traditional business products throughout the whole financial market by logically applying current information technology. This will also enable adjustments to financial institutions' organizational structures and the optimization of their offerings (Zhang, 2022). The decentralized ledger system known as the blockchain has been hyped as a game-changing technological advancement that would revolutionize several industries, including finance (Yan and Zhou, 2023). Because of blockchain, several areas, including the financial market, have evolved. Blockchain has evolved as a new paradigm for financial markets, and it is currently gaining recognition as a revolutionary technology that will impact the future.

A study identified issues with the economic and value evaluation of blockchain technology, which is related to non-financial digital assets (Yarygina et al., 2023). Blockchain is a cutting-edge technology that has the potential to transform the financial markets and is now attracting attention. Blockchain is made to safely store and use Bitcoin, a cryptocurrency. Cryptocurrency is a digital currency that functions within a decentralized blockchain system (Sidiq et al., 2022). The global economic system has changed because of cryptocurrency. It offers a fresh and creative method of exchange that has quickly impacted the conventional currency world and financial market trends (Sagheer et al., 2022). India is accelerating its journey to fully digital money in every possible way. The use of cryptocurrency by investors for financial operations including buying, selling, and trading has risen significantly in India and throughout the world (Sriman et al., 2023).

A study sought to investigate some of the distinctive characteristics of cryptocurrencies and the factors contributing to their increasing market acceptance. The goal of the study was to determine whether, in light of the growing scale of cryptocurrency markets, they might be utilized as a substitute for traditional financial assets to alleviate financial limitations faced by businesses (Gupta and Symss, 2023). One of the studies drew attention to a unique aspect of trading cryptocurrencies and provided data on investor behavior that isn't accessible in conventional financial markets (Hoang and Baur, 2022). One research suggested a novel methodology that forecasts unforeseen circumstances of speculative activities in the cryptocurrency sector (Alaminos and Belén Salas, 2023). One study successfully illustrated the connection between the financial markets, economic trends, and other indicators of volatility and uncertainty and cryptocurrency environmental attention (Wang et al., 2022).

The foundation of cryptocurrency is blockchain technology, which is theoretically decentralized—that is, not supervised by a higher authority (Ramakrishnan et al., 2023). The digital financial markets have benefited from the increasing acceptability and appeal of cryptocurrencies, but because of their decentralization and confidentiality, there has also been a rise in criminality. Sufficient oversight of decentralized money, specifically Bitcoin, can shield users from monetary losses and enhance the community (Xiang et al., 2022). The success of Bitcoin has led to the broad application of blockchain technology, especially in the financial industry. Other sectors, like the securities market and energy trading, have developed as a result of blockchain (Gorkhali and Chowdhury, 2021).

1.4 METHODOLOGY

The process of collecting data is shown in Figure 1.1:

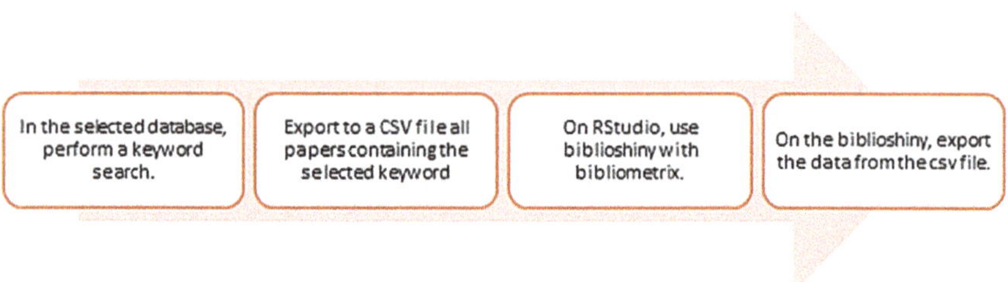

FIGURE 1.1 Data collection process.

Bibliometric techniques are a widely used review methodology that enables scholars to objectively and quantitatively evaluate a set of publications in order to identify significant similarities, connections, and patterns (Homrich et al., 2018). Frequency counts and citation patterns were used to profile the sample in order to identify the most important contributions (Kumar et al., 2020). The average number of citations for each publication and the total number of scientific publications published each year are counted before the sample analysis commences. Subsequently, an analysis was carried out to identify the journals, authors, and articles in the sample that had the greatest influence. The citation analysis and profiling were carried out using Bibliometrix, an R-based tool for mapping literature (Aria and Cuccurullo, 2017).

Using VOSviewer software, a bibliographic network was established in order to understand the key fields of research on emerging paths in financial markets using blockchain technology. Based on their bibliographic coupling, the range of references that papers have reveals how similar they are. In a system of bibliographic articles, two articles have more references the more closely they are related (Zupic and Čater, 2015). A co-occurrence analysis is used in the VOSviewer to identify how frequently the terms appear together. The co-occurrence analysis shows how frequently the keyword appears in the collected sample of research.

1.4.1 *SEARCH TERMS AND SAMPLE SELECTION*

Sample articles are sorted based on two keywords considered, which are "financial markets" and "blockchain technology," from a research database that is Scopus. The time span is from 2015–2023. This resulted in 328 samples of documents. Some basic information about the samples is presented in Figure 1.2.

Figure 1.2 shows the main information of the data. There is a total of 232 sources in 328 samples of documents. The annual growth rate is 47.16%. The total number of authors in all the documents is 861. International Co-Authorship as co-authors from other countries is 57. The co-Author per document is 2.96. The total number of references in the sample is 13090. The average citation per document is 11.1.

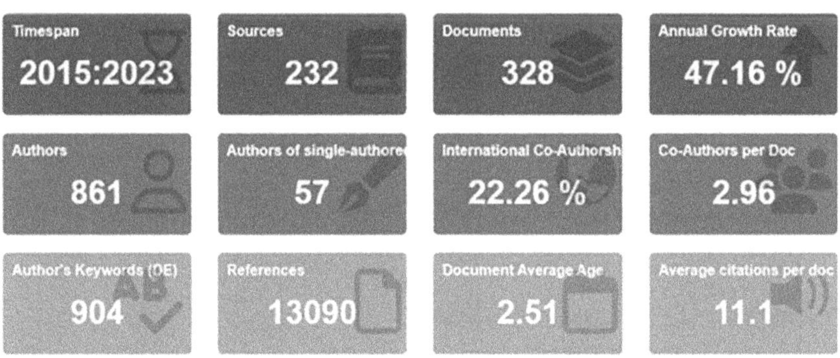

FIGURE 1.2 Basic information.

1.5 ANALYSIS

Documents are analyzed on four grounds:

- Literature profiling;

- Thematic analysis;
- Bibliographic coupling; and
- Co-occurrence analysis.

1.5.1 LITERATURE PROFILING

Literature profiling is used to show the development of sustainability reporting with the use of technology. The number of studies published in this area has been shown.

The yearly output of articles over the years is shown in Figure 1.3. The number of years was indicated by the x-axis, while the number of articles was indicated by the y-axis. Through analysis, it is evident that prior to 2015, no research was conducted on the use of technology in the financial markets. From 2016, studies on technology in financial markets started increasing and were at their maximum in 2022. In 2023, it can be seen that research has declined.

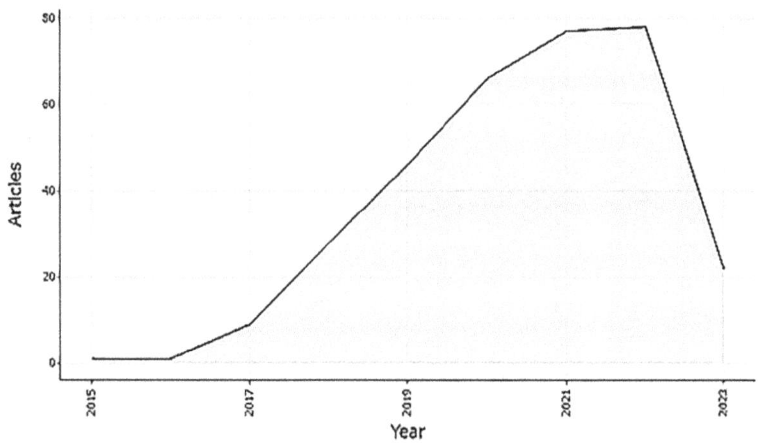

FIGURE 1.3 Annual scientific production.

The average number of citations per year is shown in Figure 1.4, where the y-axis represents the number of citations, and the x-axis represents the number of years. There were no studies before 2015. The number of citations started increasing in 2015, peaked in 2017, and then began to decline in 2018, continuing to decline until 2020. In 2020, the number of citations increased again, but after that, it continuously decreased.

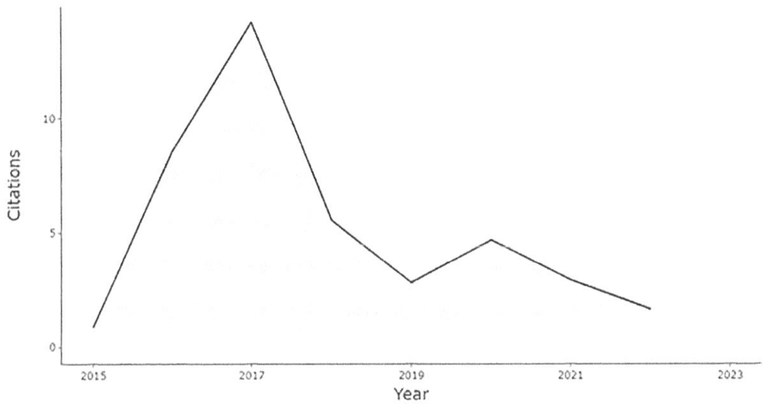

FIGURE 1.4 Average citation per year.

1.5.1.1 COUNTRY SPECIFIC ANALYSIS

Two types of charts have been produced using biblioshiny to do the country-specific analysis of research on financial markets using blockchain technology.

Figure 1.5 depicts country-specific article production over time. The x-axis indicates the number of years, while the y-axis indicates the total number of articles produced by each country. Figure 1.5 depicts the top 10 countries' production of articles from 2015–2023.

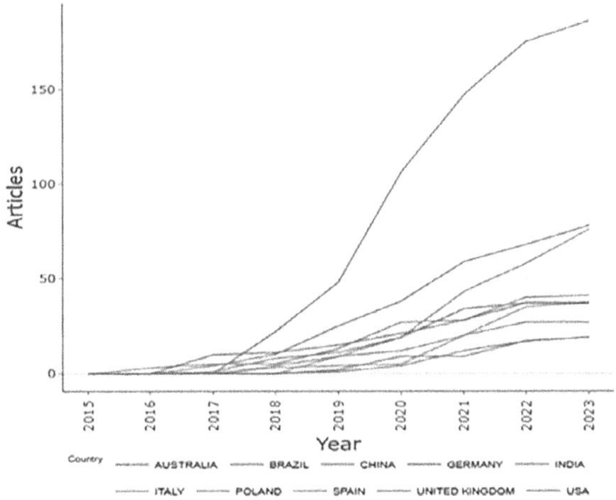

FIGURE 1.5 Country-wise production of articles.

China has 186 articles, which is the maximum number of articles by 2023, whereas the USA and India have the 2nd and 3rd highest number of articles, respectively. Then Australia has 41 articles, and Italy, Germany, and the UK have 37 articles each. Poland has 27 articles by 2023. Lastly, Brazil and Spain have both produced only 19 articles by 2023.

Figure 1.6 displays the number of citations for the top 10 countries. The x-axis shows the number of citations, while the y-axis shows the top 10 countries. The maximum citations, 529, are from the USA, whereas the lowest is from Lebanon, with 70. China has 406 citations, followed by the United Kingdom with 252, Denmark with 135, Korea with 123, Poland with 120, India with 82, Germany with 76, and Croatia with 74. From Figure 1.6, it can be interpreted that although India is producing a high number of articles, it has a lower number of citations.

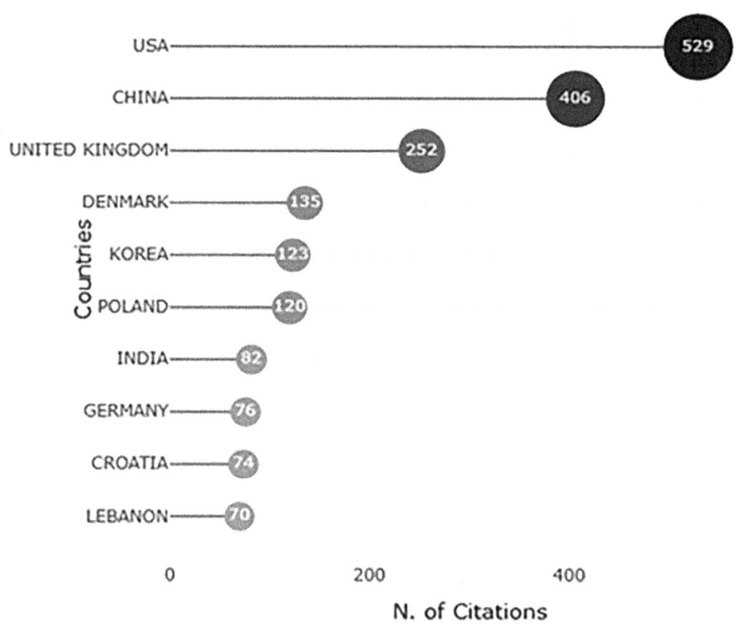

FIGURE 1.6 Top 10 cited countries.

1.5.1.2 AUTHOR ANALYSIS

The author from the sample with the most citations is presented by examining the most cited author. The top 10 cited authors are presented in Figure 1.7.

Figure 1.7 depicts the top 10 locally cited authors. On the x-axis, the total number of local citations is shown, and on the y-axis, authors are listed. Saha S, Sakhare NN, and Shaik IS all have been cited a maximum of seven times locally. Li Y has been cited six times locally. Rani R, Sable A, and Srinivas V, all cited five times locally, Chen Y, Ivanov N, and Milionis J all have been cited four times locally.

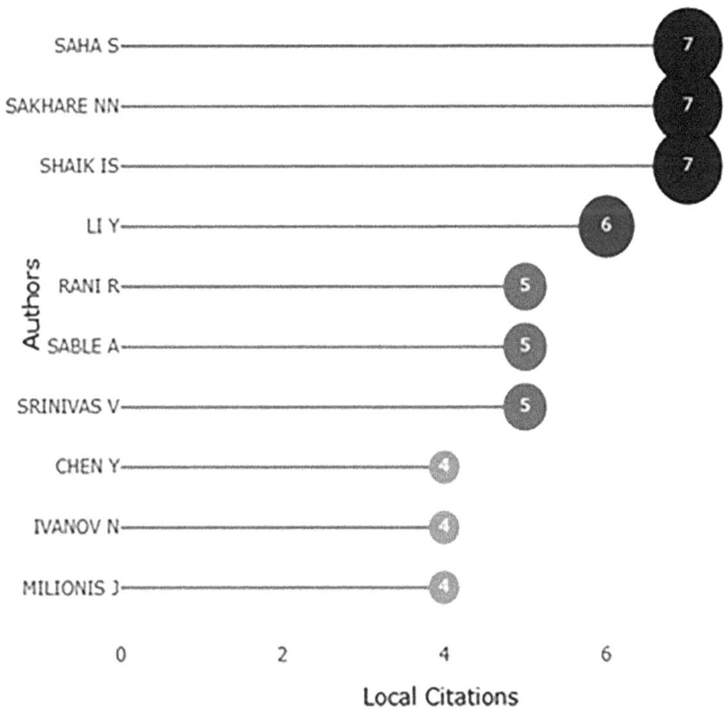

FIGURE 1.7 Top 10 locally cited authors.

1.5.1.3 ANALYSIS OF DOCUMENTS

The criteria used to assess the influence of papers are the most frequently cited documents globally. Figure 1.8 shows the top 10 documents with their global citations. A paper by Gomber provides a fintech innovation mapping technique that makes it possible to evaluate the degree of changes and transformations occurring in the financial services industry (Gomber et al., 2018) and has a maximum number of global citations of 471. The paper by Dai intends to give a basic overview of how blockchain technology might facilitate a transparent, verifiable, and real-time accounting ecosystem (Dai and Vasarhelyi, 2017) and has global citations of 300. Hayes (2017) has 202 global citations, whereas; Chang et al. (2020); Hyvärinen et al. (2017) and Yao et al. (2019) have 165, 158, and 121, respectively. Albayati et al. (2020) has 101 global citations. Both Begušić et al. (2018) and Schär (2021) have 74 global citations each. Lastly, Wątorek et al. (2021) has global citations of 72.

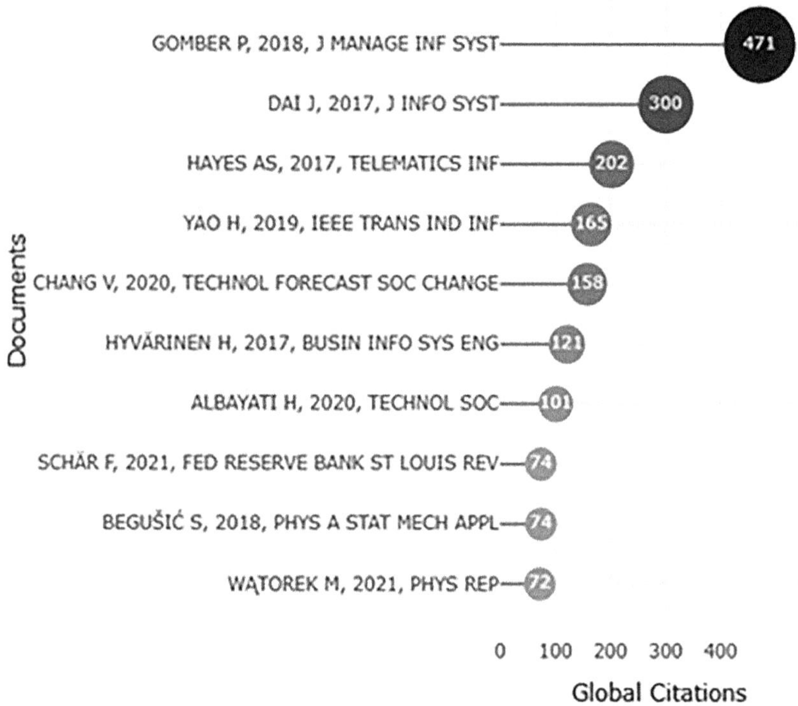

FIGURE 1.8 Top 10 globally cited documents.

1.5.1.4 ANALYSIS OF JOURNALS

The greatest quantity of publications published in a journal indicates its impact. With a maximum number of documents, Figure 1.9 displays the top 10 relevant journals.

Figure 1.9 depicts the top 10 journals with the highest number of publications. The X-axis shows the quantity of documents, while the Y-axis indicates the top 10 journals. Lecture Notes in Computer Science has the maximum number of documents, that is, 12. ACM International Conference Proceedings has 11 documents. Lecture Notes in Networks and Systems has 10 documents. Smart Innovation, Systems, and Technology, and Technological Forecasting and Social Change have seven documents. IEEE Access has five documents, whereas both Communication in Computer and Information Science, and Lecture Notes on Data Engineering and Communication have four documents. At last, Crypto Uncovered and Finance: Theory and Practice both have three documents each.

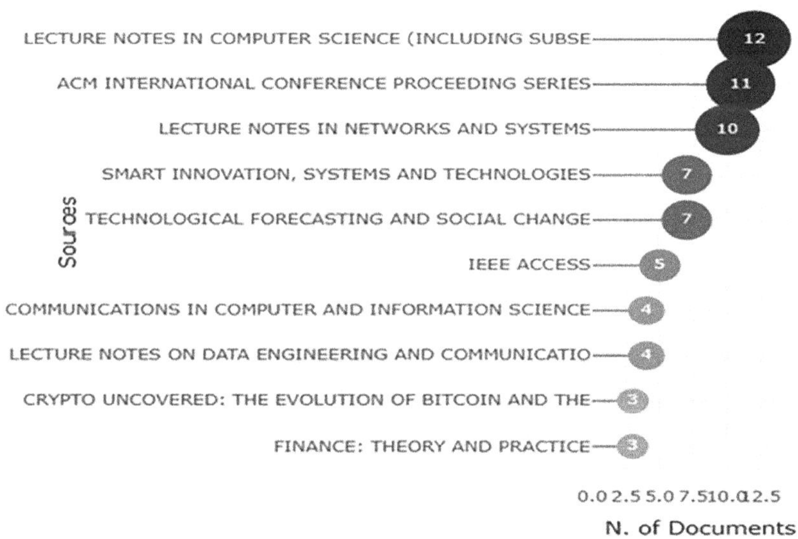

FIGURE 1.9 Top 10 journals with maximum documents.

1.5.1.5 KEYWORD ANALYSIS

Figure 1.10 depicts the top 10 keywords that occurred most frequently in the documents. The X-axis denotes numerous occurrences, while the Y-axis indicates the top 10 keywords. "Blockchain," as a keyword, has a maximum occurrence of 146. "Financial markets" is second, with 103 occurrences. "Commerce" occurred 61 times, "block-chain" occurred 53 times, and "Investments" occurred 45 times. "Bitcoin" and "electronic trading" both occurred 29 times, whereas "financial market" occurred 24 times. Lastly, both "costs" and "fintech" occurred 23 times.

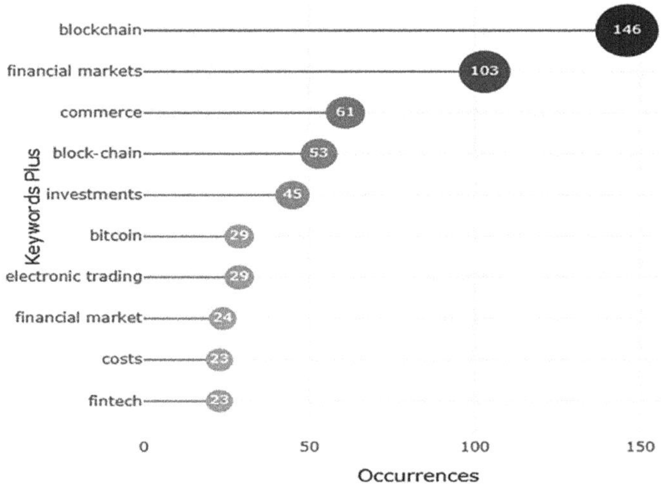

FIGURE 1.10 Keyword occurrence over the years.

1.5.2 THEMATIC ANALYSIS

On a graph, the co-occurrence network clusters were shown as bubbles. The number of times a word appears in the cluster determines the size of the bubble. The X-axis displays network cluster centrality, or the degree of connectivity between graph clusters, and denotes the significance of a study issue. The Y-axis displays density, which is a measurement of a cluster network's internal strength and theme expansion (Cahlik, 2000; Callon et al., 1991).

According to Alkhammash (2023), graphing themes showed the following:

- **Motor Themes (First Quadrant):** The cluster network's high centrality and density suggest that themes are useful for structuring study topics.
- **Niche Themes (Second Quadrant):** High density and low centrality imply that the themes are of little importance.
- **Emerging or Declining Themes (Third Quadrant):** Low centrality and low density suggest that themes are limited and undeveloped.
- **Basic Themes (Fourth Quadrant):** Low density and high centrality indicate their limited development.

Figure 1.11 shows the thematic analysis of the documents of financial markets using blockchain technology. The biggest bubble lies in the motor as well as the basic theme quadrant where keywords like blockchain and financial markets lie. The second biggest bubble lies in the basic theme quadrant with keywords like fintech, finance, and risk assessment. The niche themes quadrant includes keywords like financial reports, information disclosure, accounting information, digital assets, decentralized finance, and decentralized exchange. The emerging or declining themes quadrant include keywords like sustainable development, capital markets, and Covid-19.

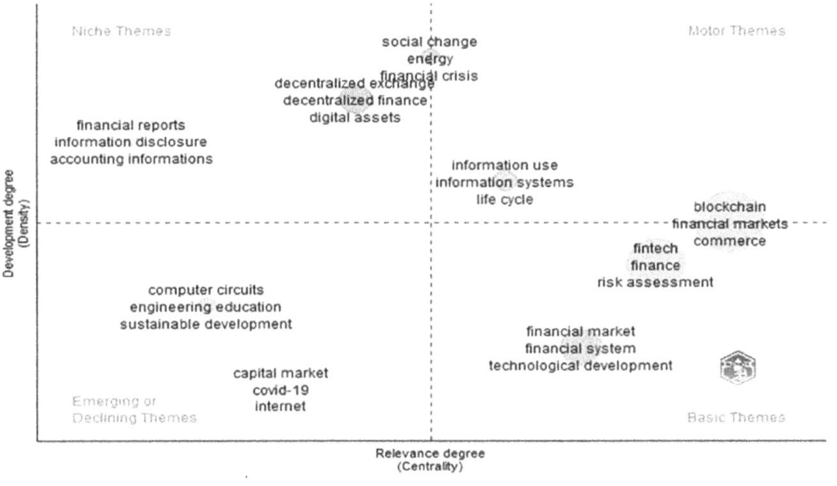

FIGURE 1.11 Thematic analysis.

1.5.3 BIBLIOGRAPHIC COUPLING

Bibliographic coupling counts the number of references that two papers have in common to determine how similar they are. Using documents as the analytical unit, Figure 1.12 illustrates bibliographic coupling (with the use of VOSviewer). The listed documents are those that have been cited in other documents. Figure 1.12 displays the author and the year of publication of the documents as circles. The number of citations for a certain text is indicated by the size of the circle; the larger the circle, the more citations it has, and vice versa. The degree of relationship between the papers is shown by the lines joining the circles. The lines indicate how many links the two documents share. Documents arranged adjacent to one another indicate a closer relationship than is the case with other documents.

The minimum number of citations per document was set to five which was only met by 104 documents out of 328. Only 85 documents out of

104 are interconnected which resulted in Figure 1.12 representing 10 clusters. 538 connections were made. According to the criteria mentioned, the highest cluster is red, followed by green, dark blue, yellow, purple, light blue, orange, brown, pink, and peach. The colors of these clusters show the total quantity of citations and research output. The documents in the sample with the highest number of citations and research output are displayed in the first cluster, which is colored red. Documents from Cluster 1 included early discussions about how blockchain technology could facilitate a transparent, verifiable, and real-time accounting ecosystem (Dai and Vasarhelyi, 2017). One of the studies in the second-highest cluster, the green color cluster, highlights the key features of FinTech and Blockchain technology and discusses their influence and revolution in the financial industry (Chang et al., 2020). The third cluster, represented by the color dark blue, seeks to identify the most likely factors that influence the construction of cryptocurrency values, including the value of Bitcoin. Since Bitcoin is becoming more and more popular and accepted by merchants, it is crucial to try to comprehend the elements that affect its value-building (Hayes, 2017). The following cluster, which is represented by the color yellow, summarizes the primary use cases for blockchain technology by giving examples and going over how and when to use them from a technical point of view (Rodrigues et al., 2018). The following cluster, called Purple, introduces a novel fintech innovation mapping technique that allows evaluation of the degree of modifications and revolutions in financial services domains (Gomber et al., 2018). The next cluster, light blue, provides an overview of blockchain technology and the history of cryptocurrencies. It has often been demonstrated that cryptocurrencies and the exchanges on which they are exchanged differ from one another (Wątorek et al., 2021). Subsequently, Cluster Orange denoted a study on a collaborative strategy utilizing both blockchain technology (BCT) and rationality in Multi-Agent Systems (MAS) decision-making, an artificial intelligence breakthrough (Calvaresi et al., 2019). The next cluster, shown by the brown color, demonstrates how the decentralized solution can provide lower transaction fees by substituting mining fees—which are meant to reward miners for their dependable work maintaining the integrity of the system—for the commissions owed to brokers and central authorities (Po et al., 2018). The following cluster, Pink, conducted a methodical analysis of blockchain technology and its use in the financial and economic sectors (Zhang et al., 2020). In the last Peach color cluster, a

conceptual framework for the management of blockchain-based networks in the financial sector is developed (Paech, 2017).

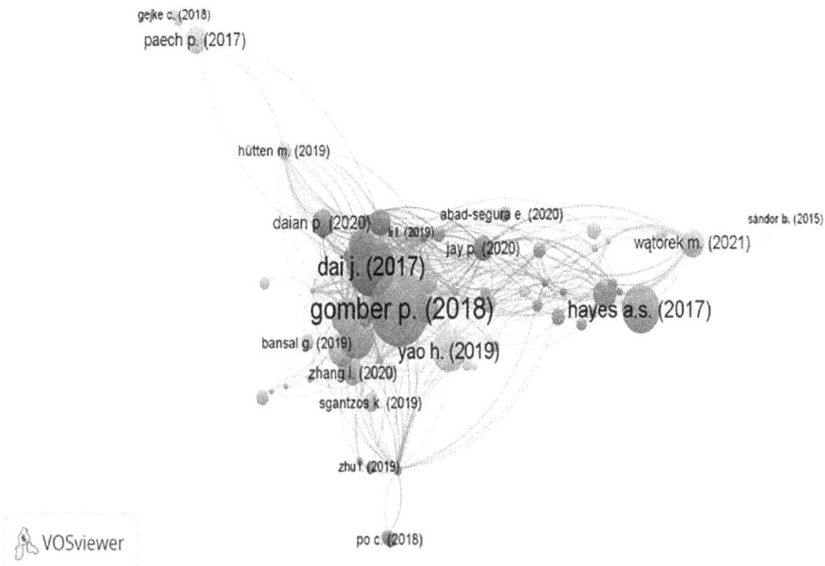

FIGURE 1.12 Bibliographic coupling.

1.5.4 CO-OCCURRENCE ANALYSIS

Co-occurrence analysis shows how closely connected two terms or keywords are to one another (Eck and Waltman, 2016). The distance on a map between two terms indicates their degree of relatedness; the closer the distance, the closer the terms are related. The full counting method had been selected as the counting method, and the co-occurrence analysis was carried out using the VOS viewer option for all keywords.

Figure 1.13 shows the co-occurrence of keywords in the papers on financial markets and blockchain technology. A minimum number of occurrences was set as 5. There were a total of 2030 keywords. The total number of keywords that showed relatedness was 76. The total number of links was 1058. It displays seven clusters in various colors. The hues of a cluster are determined by the density and similarity of the keywords contained inside it. This means that the red cluster has the maximum density, followed by the green, dark blue, yellow, purple, light blue, and orange clusters, which are the remaining clusters. Cluster red includes keywords like the financial market, fintech, blockchain technology, and innovation. Cluster green includes keywords like bitcoin, cryptocurrency, and financial system. The dark blue cluster includes keywords like financial markets, commerce, block-chain, and stock exchange. The yellow cluster includes keywords like blockchain, cryptocurrencies, and liquidity. The purple cluster includes keywords like neural networks, electronic trading, machine learning, and forecasting. The light blue cluster includes keywords like regulation, laws, and legislation, and risk assessment. Last, the orange cluster includes keywords like investments, decision making, technological innovation, and information systems.

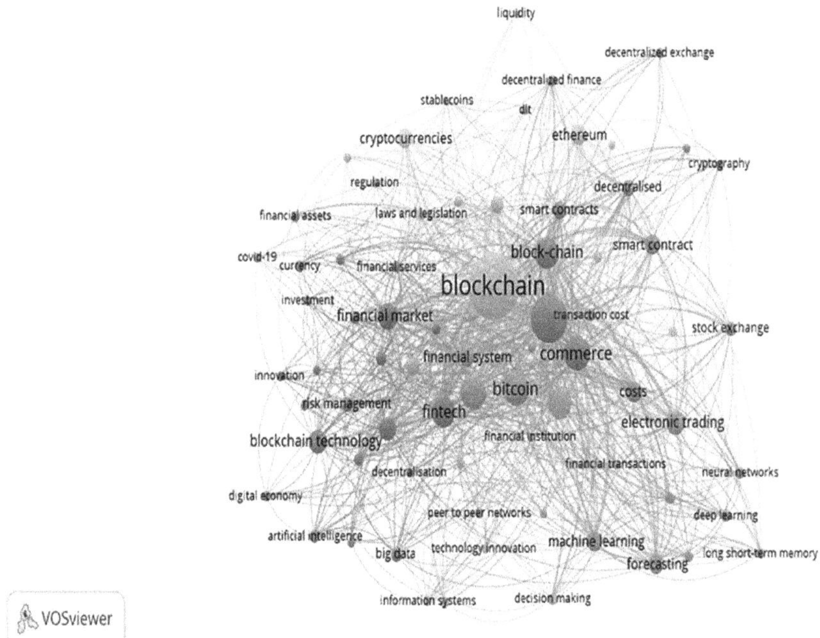

FIGURE 1.13 Co-occurrence analysis of keywords.

1.6 CONCLUSION

Technology in finance is currently in the spotlight as AI (Artificial Intelligence) and blockchain technology are gaining more interest from researchers. The research reveals the development of financial markets with the use of technology using bibliometric and co-occurrence analysis. In the profile analysis section, the timespan of this research was from 2015–2023. The total number of documents was 328, having an annual growth rate of 47.16%.

In the literature profiling phase, papers, journals, authors, and keyword occurrences were also assessed. The article additionally conducts a country-wise analysis to determine which nation is generating the most research on financial markets using blockchain technology. It was discovered that the highest number of articles are from China, whereas India is the 3rd highest. Thematic analysis showed keywords like blockchain and financial markets come under the motor themes, whereas fintech, risk assessment, and finance lie under the basic themes, which need more importance and development. Capital markets and sustainable development keywords lie under the declining or emerging themes that are currently underdeveloped and need development. Co-occurrence analysis and bibliographic coupling were carried out in the section that follows. All of the papers were subjected to bibliometric coupling, which involved developing links between the best research by utilizing various clusters. Through co-occurrence analysis, it was discovered that there was a total of 2030 keywords in the total documents, out of which 76 keywords met the threshold of five set as the number of occurrences of a keyword. The blockchain keyword occurred 146 times in the articles, whereas the Financial Market keyword occurred 103 times.

The innovation, risk, and other systems of the financial market are currently severely hindered by blockchain technology, necessitating that a high priority be placed on the innovative development of blockchain technology in the financial markets sector. As a result, a progressive path strategy needs to be adopted.

1.7 LIMITATIONS

The keywords used in this paper's searches could be selective. The inclusion of more keywords may be useful in future research.

A minimum number of published papers should be considered while using the search engines for the study. Other search engines besides Scopus may be used by researchers in the future.

KEYWORDS

- **blockchain technology**
- **financial markets**
- **thematic analysis**
- **bibliographic analysis**
- **co-occurrence analysis**
- **literature profiling**

REFERENCES

Alaminos, D., & Belén Salas, M. (2023). Multiscale multifractal detrended analysis of speculative attacks dynamics in cryptocurrencies. *Lecture Notes in Computer Science (including subseries Lecture Notes in Artificial Intelligence and Lecture Notes in Bioinformatics)*, *13588*, 325–339. https://doi.org/10.1007/978-3-031-23492-7_28/COVER (accessed on 26th July 2025).

Albayati, H., Kim, S. K., & Rho, J. J. (2020). Accepting financial transactions using blockchain technology and cryptocurrency: A customer perspective approach. *Technology in Society*, *62*, 101320. https://doi.org/10.1016/J.TECHSOC.2020.101320 (accessed on 26th July 2025).

Alkhammash, R. (2023). Bibliometric, network, and thematic mapping analyses of metaphor and discourse in COVID-19 publications from 2020 to 2022. *Frontiers in Psychology*, *13*. https://doi.org/10.3389/FPSYG.2022.1062943/FULL (accessed on 26th July 2025).

Aria, M., & Cuccurullo, C. (2017). bibliometrix: An R-tool for comprehensive science mapping analysis. *Journal of Informetrics*, *11*(4), 959–975. https://doi.org/10.1016/j.joi.2017.08.007 (accessed on 26th July 2025).

Begušić, S., Kostanjčar, Z., Eugene Stanley, H., & Podobnik, B. (2018). Scaling properties of extreme price fluctuations in Bitcoin markets. *Physica A: Statistical Mechanics and Its Applications*, *510*, 400–406. https://doi.org/10.1016/J.PHYSA.2018.06.131 (accessed on 26th July 2025).

Cahlik, T. (2000). Comparison of the maps of science. *Scientometrics*, *49*(3), 373–387.

Callon, M., Courtial, J. P., & Laville, F. (1991). Co-word analysis as a tool for describing the network of interactions between basic and technological research: The case of polymer chemsitry. *Scientometrics*, *22*(1), 155–205. https://doi.org/10.1007/BF02019280 (accessed on 26th July 2025).

Calvaresi, D., Mualla, Y., Najjar, A., Galland, S., & Schumacher, M. (2019). Explainable multi-agent systems through blockchain technology. *Lecture Notes in Computer Science*, *11763*, 41–58. https://doi.org/10.1007/978-3-030-30391-4_3 (accessed on 26th July 2025).

Chang, V., Baudier, P., Zhang, H., Xu, Q., Zhang, J., & Arami, M. (2020). How blockchain can impact financial services–The overview, challenges and recommendations from expert interviewees. *Technological Forecasting and Social Change*, *158*, 120166. https://doi.org/10.1016/J.TECHFORE.2020.120166 (accessed on 26th July 2025).

Dai, J., & Vasarhelyi, M. A. (2017). Toward blockchain-based accounting and assurance. *Journal of Information Systems*, *31*(3), 5–21. https://doi.org/10.2308/ISYS-51804 (accessed on 26th July 2025).

Eck, N. J. van, & Waltman, L. (2016). Text mining and visualization. *Text Mining and Visualization*, 1–5. https://doi.org/10.1201/b19007 (accessed on 26th July 2025).

Gomber, P., Kauffman, R. J., Parker, C., & Weber, B. W. (2018). On the fintech revolution: Interpreting the forces of innovation, disruption, and transformation in financial services. *Journal of Management Information Systems*, *35*(1), 220–265. https://doi.org/10.1080/07421222.2018.1440766 (accessed on 26th July 2025).

Gorkhali, A., & Chowdhury, R. (2021). Blockchain and the evolving financial market: A literature review. *Journal of Industrial Integration and Management*, *7*(1), 47–81. https://doi.org/10.1142/S242486222150024X (accessed on 26th July 2025).

Gupta, N., & Symss, J. (2023). Can cryptocurrency solve the problem of financial constraint incorporates? A literature review and theoretical perspective. *Qualitative Research in Financial Markets*. https://doi.org/10.1108/QRFM-12-2021-0215/FULL/XML (accessed on 26th July 2025).

Hayes, A. S. (2017). Cryptocurrency value formation: An empirical study leading to a cost of production model for valuing bitcoin. *Telematics and Informatics*, *34*(7), 1308–1321. https://doi.org/10.1016/J.TELE.2016.05.005 (accessed on 26th July 2025).

Hoang, L. T., & Baur, D. G. (2022). Loaded for bear: Bitcoin private wallets, exchange reserves and prices. *Journal of Banking & Finance*, *144*, 106622. https://doi.org/10.1016/J.JBANKFIN.2022.106622 (accessed on 26th July 2025).

Homrich, A. S., Galvão, G., Abadia, L. G., & Carvalho, M. M. (2018). The circular economy umbrella: Trends and gaps on integrating pathways. *Journal of Cleaner Production*, *175*, 525–543. https://doi.org/10.1016/j.jclepro.2017.11.064 (accessed on 26th July 2025).

Hyvärinen, H., Risius, M., & Friis, G. (2017). A blockchain-based approach towards overcoming financial fraud in public sector services. *Business and Information Systems Engineering*, *59*(6), 441–456. https://doi.org/10.1007/S12599-017-0502-4/METRICS (accessed on 26th July 2025).

Kumar, B., Sharma, A., Vatavwala, S., & Kumar, P. (2020). Digital mediation in business-to-business marketing: A bibliometric analysis. *Industrial Marketing Management*, *85*(May), 126–140. https://doi.org/10.1016/j.indmarman.2019.10.002 (accessed on 26th July 2025).

Liu, T., & Yu, Z. (2022). The analysis of financial market risk based on machine learning and particle swarm optimization algorithm. *EURASIP Journal on Wireless Communications and Networking*, *2022*(1), 1–17. https://doi.org/10.1186/S13638-022-02117-3/FIGURES/14 (accessed on 26th July 2025).

Paech, P. (2017). The governance of blockchain financial networks. *The Modern Law Review*, *80*(6), 1073–1110. https://doi.org/10.1111/1468-2230.12303 (accessed on 26th July 2025).

Po, C., Po, C., Marce, A., Ves, A., Petrica, T., Cioar, T., Anghe, I., & Salomi, I. (2018). Decentralizing the stock exchange using blockchain: An Ethereum-based implementation of the Bucharest Stock Exchange. *Proceedings – 2018 IEEE 14th International Conference on Intelligent Computer Communication and Processing, ICCP 2018*, 459–466. https://doi.org/10.1109/ICCP.2018.8516610 (accessed on 26th July 2025).

Ramakrishnan, R., Vadakedath, A., Bhaskar, A., Sachin Kumar, S., & Soman, K. P. (2023). Data-driven volatile cryptocurrency price forecasting via variational mode decomposition and BiLSTM. *Lecture Notes in Networks and Systems*, *473*, 651–663. https://doi.org/10.1007/978-981-19-2821-5_55/COVER (accessed on 26th July 2025).

Rodrigues, B., Bocek, T., & Stiller, B. (2018). The use of blockchains: Application-driven analysis of applicability. *Advances in Computers*, *111*, 163–198. https://doi.org/10.1016/BS.ADCOM.2018.03.011 (accessed on 26th July 2025).

Sagheer, N., Khan, K. I., Fahd, S., Mahmood, S., Rashid, T., & Jamil, H. (2022). Factors affecting adaptability of cryptocurrency: An application of technology acceptance model. *Frontiers in Psychology*, *13*, 903473. https://doi.org/10.3389/FPSYG.2022.903473/BIBTEX (accessed on 26th July 2025).

Schär, F. (2021). Decentralized finance: On blockchain- and smart contract-based financial markets. *Federal Reserve Bank of St. Louis Review*, *103*(2), 153–174. https://doi.org/10.20955/R.103.153-74 (accessed on 26th July 2025).

Sidiq, M. I., Roth, W. S., & Rusmanto, T. (2022). Literature study on the effect of blockchain usage on cryptocurrencies: Looks at comparisons of previous research. *ACM International Conference Proceeding Series*, 25–30. https://doi.org/10.1145/3549823.3549828 (accessed on 26th July 2025).

Sriman, B., Iniyal, T. T., Thasmiya, J., Ziyaadh, J. T., Ranjani, S., & Thomas, S. T. (2023). Price prediction of Bitcoin and Ethereum–A machine learning approach. *Proceedings of the International Conference on Artificial Intelligence and Knowledge Discovery in Concurrent Engineering (ICECONF)*, 1–6. https://doi.org/10.1109/ICECONF57129.2023.10084003 (accessed on 26th July 2025).

Wang, Y., Lucey, B., Vigne, S. A., & Yarovaya, L. (2022). An index of cryptocurrency environmental attention (ICEA). *China Finance Review International*, *12*(3), 378–414. https://doi.org/10.1108/CFRI-09-2021-0191/FULL/PDF (accessed on 26th July 2025).

Wątorek, M., Drożdż, S., Kwapień, J., Minati, L., Oświęcimka, P., & Stanuszek, M. (2021). Multiscale characteristics of the emerging global cryptocurrency market. *Physics Reports*, *901*, 1–82. https://doi.org/10.1016/J.PHYSREP.2020.10.005 (accessed on 26th July 2025).

Xiang, Y., Li, T., & Li, Y. (2022). Leveraging subgraph structure for exploration and analysis of Bitcoin address. *Proceedings–2022 IEEE International Conference on Big Data, Big Data*, 1957–1962. https://doi.org/10.1109/BIGDATA55660.2022.10020980 (accessed on 26th July 2025).

Yan, W., & Zhou, W. (2023). Is blockchain a cure for peer-to-peer lending? *Annals of Operations Research*, *321*(1), 693–716. https://doi.org/10.1007/S10479-022-05108-1/FIGURES/7 (accessed on 26th July 2025).

Yao, H., Mai, T., Wang, J., Ji, Z., Jiang, C., & Qian, Y. (2019). Resource trading in blockchain-based industrial internet of things. *IEEE Transactions on Industrial Informatics*, *15*(6), 3602–3609. https://doi.org/10.1109/TII.2019.2902563 (accessed on 26th July 2025).

Yarygina, I. Z., Pomulev, A. A., & Bakhramov, A. A. (2023). Blockchain as an object of value assessment. *Lecture Notes in Networks and Systems*, *596*, 205–215. https://doi.org/10.1007/978-3-031-21435-6_17/COVER (accessed on 26th July 2025).

Zhang, L., Xie, Y., Zheng, Y., Xue, W., Zheng, X., & Xu, X. (2020). The challenges and countermeasures of blockchain in finance and economics. *Systems Research and Behavioral Science*, *37*(4), 691–698. https://doi.org/10.1002/SRES.2710 (accessed on 26th July 2025).

Zhang, W. (2022). The current situation and trend of blockchain technology in the financial field. *Mobile Information Systems*, *2022*. https://doi.org/10.1155/2022/7908625 (accessed on 26th July 2025).

Zupic, I., & Čater, T. (2015). Bibliometric methods in management and organization. *Organizational Research Methods*, *18*(3), 429–472. https://doi.org/10.1177/1094428114562629 (accessed on 26th July 2025).

CHAPTER 2

From Cryptocurrency to Smart Contracts: A Blockchain's Growing Impact on Trading and Business

SHIVI MITTAL,[1,*] ANSHIKA RAJVANSHI,[2] and ANKUR PANDEY[3]

[1]*Mangalmay Institute of Management & Technology, Greater Noida, Uttar Pradesh, India*

[2]*IMS, Ghaziabad, Uttar Pradesh, India*

[3]*Aadhar Housing Finance Ltd. Mumbai, Maharashtra, India*

*Corresponding author

ABSTRACT

In the previous research of the authors, the dynamics of cryptocurrency using blockchain technology have been studied. The chapter captures the present state of research on legal challenges related to the applicability of cryptocurrency in India by providing a critical review. An overview of pre- and post-pandemic transactions by investors in digital currency has been discussed and reviewed. In the current study, the author(s) try to examine the impact of blockchain technology on trading and business, with an emphasis on the growth and sustainability of the business. The business process will benefit from effective tracking, visibility, security improvements, and cost savings as a result (Pal et al., 2021). Therefore, to ensure the legitimacy of such items, trust and confidence are factors that need to be considered (Loebbecke and Lueneborg, 2018). Through a systematic review of the literature, the application in various aspects of different types of businesses is explored, identifying the challenges in

FinTech and Blockchain Technology in Financial Markets. Neha Puri, Anjali Munde, Emilia Alaverdov (Eds.)
© 2027 Apple Academic Press, Inc. Co-published with CRC Press (Taylor & Francis)

DOI: 10.1201/9781779641298-2

blockchain implementation and looking for future trends along with the regulatory framework of trading and business in India. This chapter is important for scholars, researchers, and even entrepreneurs to understand the pedagogy behind using any technology with safe and secure transactions in business.

2.1 INTRODUCTION

Blockchain was first used in the banking sector for cryptocurrencies so that digital authentications can take the place of manual transaction authentication. Since blockchain could improve data security, immutability, trust, traceability, and transparency, it has drawn interest from academics and industry people. Investment in blockchain technology is increasing quickly, and by 2025, it is projected to reach USD 176 billion (Amstelveen, 2018). As previously mentioned, blockchain technology has emerged as a decentralized computing paradigm that effectively addresses the issue of trust associated with a centralized authority. In a blockchain network, multiple nodes work together to ensure and preserve a shared set of transaction records in a distributed manner without the need for any trusted entity. Within the network, certain nodes called miners are tasked with appending new blocks to a widely accessible ledger known as the blockchain. Blockchain Technology (BCT) has three vital principles which are important to mention here.

Decentralization → Transparency → Security

Blockchain, a distributed system, is designed to bring transparency to various aspects of our lives. It has the potential to revolutionize digital transactions and even decentralize resources like water. This system aims to address global issues in a fair and ethical manner, reshaping the world and its problem-solving strategies. The key strength of blockchain lies in its decentralization, as it eliminates the need for a central authority. Through simple yet effective algorithms, this technology effectively tackles the problem of centralization. Blockchain has the capacity to redefine how businesses operate, how governments provide services, and how social work is conducted. However, despite being introduced years ago, there remains skepticism surrounding its adoption by many individuals and businesses (Sudhakaran et al., 2019). Through this chapter, we discuss various case studies that show how the business is evolving by using BCT. The application of blockchain technology has extended from digital currency into finance, and it has even gradually extended into health care, supply chain management, market monitoring, smart energy, and copyright protection (Engelhardt, 2017; Laskowski, 2018). Through this chapter, we try to examine the impact of Blockchain technology on the sustainability of businesses.

The following are a few ways that blockchain technology is being used in business and trade:

1. **Payment Systems:** Payment systems without middlemen, like banks or payment processors, can be made secure and quick using blockchain technology. This could aid in lowering transaction fees, speeding up transactions, and enhancing payment transaction security.

2. **Smart Contracts:** In smart contracts, the terms of the agreement between the seller and the buyer are expressly written into the code. These commitments are met on their own. Smart contracts, which can be used to schedule payments automatically under specific conditions that are satisfied, are possible to do out on a secure and open platform because of blockchain technology.

3. **Micropayments:** Small amounts of money can be exchanged using micropayments, which are made feasible by blockchain technology. This can be useful in situations when accepting standard payment methods is not feasible, such as in online gaming or content monetization.

4. **Digital Identity Verification:** Blockchain technology can be used to create safe digital identities for individuals as well as businesses. This can help to ensure that only parties who have been verified are involved in transactions and to prevent identity fraud.

Blockchain technology can greatly improve payment systems by providing a fast, safe, and efficient platform for transactions. It is a desirable choice for both individuals and companies because it does away with the need for middlemen and lowers transaction costs. Before blockchain-based payment systems are extensively used, there are still issues that need to be solved, like scalability and regulatory frameworks.

2.2 IMPACT OF BLOCKCHAIN TECHNOLOGY ON TRADING AND BUSINESS SUSTAINABILITY

Blockchain innovation has changed exchanging and business manageability in significant ways. Its inborn straightforwardness and decentralized nature foster trust among participants, dispensing with the requirement for go-betweens and reducing expenses. Transactions have gotten comfortable constantly, contrasted with the days expected in traditional frameworks, leading to faster access to funds and reduced risk (Cole and Rossana, 2019). Smart Contracts automate complex agreements, reducing errors and administrative overhead. This global accessibility of blockchain has enabled inclusive participation in exchanging from regions with limited access to traditional financial services. Additionally, its capacity to track and verify product authenticity has transformed the supply chain. The innovation's enhanced security features mitigate fraud and cybersecurity risks, adding to more sustainable strategic policies. Tokenization of assets on the blockchain allows for fractional ownership and trading of traditionally illiquid assets. Be that as it may, challenges, for example, administrative issues and adaptability should be addressed for more extensive adoption. Basically, blockchain's effect on exchanging and business manageability is diverse, offering efficiency, straightforwardness, security, and the potential for new plans of action. In total, the impact of BCT on trading and business sustainability is multifaceted. The part concludes with an investigation of the blockchain concept insinuating its effectiveness from one viewpoint and its functional drawbacks on the other.

In the areas of trade, transportation, logistics, and supply chain management, the application of blockchain technology presents a huge possibility for streamlining procedures and boosting business models. As trade is a crucial element for our economic viability, it requires huge procedures to execute it right from connecting people, operating with various activities, handling and exchanging data and documents and so forth (Al Barghuthi et al., 2018).

2.2.1 BUSINESS APPLICATIONS AND SUSTAINABILITY

Numerous industries in India and overseas have implemented blockchain technology, and many of these use cases are regarded as sustainable because of their ability to increase efficiency, security, and transparency while lowering fraud and waste. Some examples of sectors and companies using blockchain technology are as follows:

1. **Cryptocurrencies and Exchanges:** The existence and operation of cryptocurrencies like Bitcoin and Ethereum are dependent on blockchain technology. Exchanges for cryptocurrencies make it easier to trade digital assets (Berentsen, 2019).

2. **Supply Chain Management:** Businesses are using blockchain to track the production cycle from raw materials to customers. This increases transparency and empowers consumers to make informed decisions about sustainable products (Khanfar, 2021).

3. **Food Safety:** Blockchain technology is being used by the food sector to trace the origin of food goods and ensure that they adhere to safety regulations and are untainted. It's amazing how Pietro Coricelli creates its own blockchain platform while using IBM Supply Chain Intelligence Suite to guarantee food quality, sustainable sourcing, and supply chain transparency. The company's future is being driven by the new Coricelli generation, who views technology as a crucial tool for ensuring transparency, sustainability, and high standards of quality (IBM Food Trust and Coricelli, n.d.).

4. **Healthcare:** Patient records are securely shared and stored via blockchain, protecting the privacy and integrity of the data. Clinical trials and medication traceability can both benefit from it (Srivastava et al., 2022).

Not only in the above-mentioned areas but real estate, voting systems, intellectual property, insurance organizations, and even in making cross-border payments, blockchain technology is showing great significance. It enhances the process from the initial level to the intermediary and then to the final one. Blockchain streamlines real estate transactions by minimizing fraud and providing a tamper-proof record of ownership. In order to prevent election fraud, certain nations are investigating the use of blockchain for safe and transparent voting systems. Blockchain technology enables writers, artists, and inventors to timestamp and safeguard their intellectual property rights. Blockchain can improve claim processing accuracy and streamline insurance operations while lowering fraud. Blockchain makes cross-border payments easier and faster, which benefits international trade and remittance services.

2.3 CHALLENGES AND OPPORTUNITIES

A lot of problems need to be overcome for blockchain technology to become more widely adopted, scalable, and efficient. Here are a few of the main obstacles:

1. **Scalability:** One of the main issues facing blockchain technology at the moment is scalability. Large-scale applications cannot be served by current blockchain systems due to their limited capacity for processing transactions. Several approaches, such as sharing and layer-two scaling strategies like Lightning Network, are being investigated by developers as potential answers to this issue.
2. **Interoperability:** One further significant problem is the incompatibility of different blockchain networks. It is now difficult for different blockchain networks to communicate with one another because they operate independently. As a result, creating applications that work flawlessly across several blockchains becomes challenging.
3. **Security:** While assaults are rare with blockchain technology, they are nevertheless feasible. Hackers have been known to exploit the flaws in smart contracts and decentralized applications (dApps) to steal money or interfere with operations. The security of blockchain networks needs to be strengthened in order to maintain customer trust.
4. **Regulation:** There are benefits and drawbacks to the fully unregulated environment in which blockchain technology operates. Although it encourages innovation and experimentation, it also creates uncertainty and can make it difficult for businesses to operate legally. As blockchain technology is utilized more widely, governments and regulators will need to establish detailed laws and regulations to guarantee its correct usage.
5. **Energy Consumption:** On certain blockchains, the transaction verification mining process can consume a significant amount of energy. As a result, concerns have been raised over the potential

environmental effects of blockchain technology and the necessity for energy-efficient consensus procedures.

6. **User Adoption:** Blockchain technology is still relatively new and undiscovered by the general public, despite its potential. If developers want their products to be used more widely, they must make user-friendly user interfaces and programs that address real-world problems and offer distinct benefits.

In common parlance, if blockchain technology is to realize its full potential and be applied to transform industries and create new business opportunities, it must overcome these challenges. International application of this technology offers the highest level of transparency in the movement of goods from the producer to the end consumer. It not only helps in reducing costs but also cuts the inefficiencies of supply chain management. The unchangeable structure of the blocks eliminates fraud, and the blockchain technology offers a possible means of exchanging financial and non-financial data without the need for middlemen. By providing the essential transparency, this technology advancement may be able to address operational risk, dramatically lowering operational expenses for all stakeholders (Al Barghuthi et al., 2018).

2.4 GOVERNMENT INTERVENTION AND REGULATORY CHALLENGES IN THE ADOPTION OF BLOCKCHAIN TECHNOLOGY IN TRADING AND BUSINESS

The use of blockchain technology in trading and business is heavily influenced by governmental interference and regulatory issues. Significant interest has been generated by blockchain's ability to transform sectors thanks to its transparent and immutable ledger. Governments around the world are debating how to successfully control this disruptive technology, though.

One of the key challenges is finding a balance between promoting innovation and upholding consumer protection. For example, concerns have been raised about the illicit activities and financial stability of cryptocurrencies, which are often built on blockchain technology. Consequently, governments are implementing stringent Know Your Customer (KYC) and Anti-Money Laundering (AML) policies, such as the European Union's AMLD5 rule and the United States' FinCEN recommendations.

Cross-border impacts pose additional challenges. Blockchain's decentralized structure challenges established regulatory structures. Nations must cooperate to harmonize laws in order to guarantee seamless international trade and investment while lowering risks. In order to ensure revenue collection and prevent tax evasion, the government is also involved in tax legislation pertaining to blockchain assets, such as cryptocurrencies. These regulations require certain guidelines.

The potential benefits of blockchain technology, such as its capacity to reduce fraud, boost efficiency, and promote transparency, are known to governments. However, they are worried about the risks that can materialize, like tax evasion, money laundering, and the unchecked growth of cryptocurrencies.

2.5 REGULATORY CHALLENGES

1. **Lack of Uniformity:** Although blockchain technology is used globally, there are significant differences in national regulatory systems. Due to this irregularity, businesses and investors that conduct business worldwide encounter uncertainty.

2. **Legal Recognition:** In many areas, blockchain-based contracts and transactions are still not fully recognized by the law, which could result in disputes and other legal problems.

3. **Protection of Consumers:** With the growing use of blockchain technology, it is critical to safeguard consumers. Since there are no consumer protection laws tailored especially for blockchain technology, users are vulnerable to fraud and scams.

4. **Data Privacy:** Since blockchain technology is decentralized, information is stored across multiple nodes. Ensuring compliance with data protection legislation such as the GDPR presents a significant challenge.

5. **Tokenization:** On blockchain systems, asset and security tokenization has the power to transform the way financial markets function. However, regulating bodies are still at odds on how to classify and manage these digital assets.

6. **Cross-Border Transactions:** Blockchain facilitates cross-border transactions, but the efficiency gains that the technology had promised could be lost if regulatory barriers make these transactions more difficult or take longer.

2.6 PROSPECTIVE PATTERNS, RAMIFICATIONS, AND SUGGESTIONS ON THE INTEGRATION OF BLOCKCHAIN TECHNOLOGY IN COMMERCE AND TRADING

The direction of innovation in the previously stated industries is being influenced by the legal framework and anticipated rates of adoption of BCT in trading and commerce. While governments try to find a way to support innovation while also addressing concerns like data privacy, the methods in which they have reacted to BCT have varied in terms of security and financial stability. This means creating regulations for tokens and cryptocurrencies, altering securities frameworks for blockchain-based assets, and striking a balance between data privacy laws and blockchain transparency. Several key trends are predicted to influence the adoption of BCT in the future. As legislative clarity increases and more significant businesses look into the potential uses of blockchain, its mainstream integration into corporate processes is anticipated to increase. Scalability and cross-chain interoperability solutions will be essential for improved transaction capacity and easy communication across different blockchain networks. Blockchain adoption will also be aided by the continuous growth of supply chain traceability systems and central bank digital currency (CBDCs). The development of decentralized finance (DeFi) and environmental issues with energy-hungry proof-of-work blockchains, however, are also significant factors that will determine the course of the technology. The stage is set for a revolutionary era of blockchain implementation as industries cooperate through consortia and hybrid blockchain solutions. International trade could be revolutionized, reinvented, or disrupted by blockchain technology. This is taking place at a time when we are just starting to comprehend the effects that digitization driven by the internet has had on trade and the economy. ICT has facilitated digitization, global value chains, electronic platforms, 3D printing, and many other developments by reducing the cost of communications. It has given businesses that couldn't previously access the global market a wide range of new prospects. The internet removed entry barriers, cut distances between locations, and reduced transaction costs in that market. Smaller actors, like consumers and SMEs, can now conduct business in marketplaces that were previously unreachable. New institutions that improved trust using ICT payment systems (like PayPal, Alipay, and Klarna) emerged to facilitate this trade. Blockchain technology can replace many of these processes, automating many administrative chain functions and improving the speed, efficiency, security, and transparency of transactions. The IoT, AI, and smart contracts are just a few examples of additional emerging technologies that can be combined with blockchain to create new applications that operate similarly.

KEYWORDS

- **blockchain technology**
- **business**
- **cryptocurrency**
- **payment systems**
- **smart contracts**
- **sustainability**
- **trading**

REFERENCES

Al Barghuthi, N. B., Mohamed, H. J., & Said, H. E. (2018). Blockchain in supply chain trading. In *2018 Fifth HCT Information Technology Trends (ITT)* (pp. 336–341). IEEE.

Berentsen, A. (2019). Aleksander Berentsen recommends "Bitcoin: A peer-to-peer electronic cash system" by Satoshi Nakamoto. *21st Century Economics: Economic Ideas You Should Read and Remember*, 7–8.

Cole, R., Stevenson, M., & Aitken, J. (2019). Blockchain technology: Implications for operations and supply chain management. *Supply Chain Management: An International Journal*, *24*(4), 469–483.

Engelhardt, M. A. (2017). Hitching healthcare to the chain: An introduction to blockchain technology in the healthcare sector. *Technology Innovation Management Review*, *7*(10).

IBM. (n.d.). *IBM Food Trust and Coricelli: The Blockchain to Increase Consumer Trust. IBM Supply Chain Intelligence Suite: Food Trust: Case Study; Coricelli Italian Partners*. https://www.ibm.com (accessed on 26th July 2025).

Khanfar, A. A., Iranmanesh, M., Ghobakhloo, M., Senali, M. G., & Fathi, M. (2021). Applications of blockchain technology in sustainable manufacturing and supply chain management: A systematic review. *Sustainability*, *13*(14), 7870.

Kim, H. M., & Laskowski, M. (2018). Toward an ontology-driven blockchain design for supply-chain provenance. *Intelligent Systems in Accounting, Finance and Management*, *25*(1), 18–27.

KPMG, L. (2018). *Blockchain and the Future of Finance: A Potential New World for CFOs—And How to Prepare*. KPMG: Amstelveen, The Netherlands.

Loebbecke, C., Lueneborg, L., & Niederle, D. (2018). Blockchain technology impacting the role of trust in transactions: Reflections in the case of trading diamonds.

Pal, A., Tiwari, C. K., & Haldar, N. (2021). Blockchain for business management: Applications, challenges and potentials. *The Journal of High Technology Management Research*, *32*(2), 100414.

Srivastava, S., Pant, M., Jauhar, S. K., & Nagar, A. K. (2022). Analyzing the prospects of blockchain in healthcare industry. *Computational and Mathematical Methods in Medicine*, *2022*.

Sudhakaran, S., Kumar, S., Ranjan, P., & Tripathy, M. R. (2020). Blockchain-based transparent and secure decentralized algorithm. In *International Conference on Intelligent Computing and Smart Communication 2019: Proceedings of ICSC 2019* (pp. 327–336). Springer Singapore.

CHAPTER 3

Mapping FinTech and Blockchain Technology Research: A Bibliometric Analysis

KAJAL PANWAR,[1] NEHA PURI,[1,*] and DEEPAK TANDON[2]

[1]*Amity College of Commerce & Finance (ACCF), Amity University, Noida, Uttar Pradesh, India*

[2]*Institute of Management Studies (IMS), Ghaziabad, Uttar Pradesh, India*

*Corresponding author

ABSTRACT

Fintech, or financial technology, is undoubtedly among the most noteworthy recent breakthroughs in the world of the financial sector. As Fintech becomes more glorified, more research articles are written. It is extremely important to evaluate the caliber of these many study articles and collect relevant information. Therefore, a bibliometric examination of these articles can provide information about current research fads and pertinent issues. The Web of Science database was searched for Fintech-related publications from throughout the world that were published between 2016 and 2022. The study manages to retrieve 453 documents for analysis. Both qualitative and quantitative research have been carried out for this project. PRISMA were utilized by the authors to gather the data. Furthermore, the bibliometric analysis of the articles was conducted by VOSviewer and R-studio. A large increase in the number of international Fintech publications is predicted by the current growth patterns. The People's Republic of China made the most notable contribution to this important field. Poland, Ireland, and France are still trailing in this

FinTech and Blockchain Technology in Financial Markets. Neha Puri, Anjali Munde, Emilia Alaverdov (Eds.)
© 2027 Apple Academic Press, Inc. Co-published with CRC Press (Taylor & Francis)

DOI: 10.1201/9781779641298-3

field of study. The study's findings emphasize the key authors, institutions, nations, journals, and papers in the Fintech literature. There have also been recommendations for the Fintech literature is likely future course.

3.1 INTRODUCTION

Innovations in technology are blossoming and transforming the globe. With the availability of the internet, a wide spectrum of individuals now have easy access to lightning-fast technological breakthroughs. Industry 4.0 (Bai et al., 2020; Brettel et al., 2014; Koesworo et al., 2019; Saucedo-Martínez et al., 2018) refers to substantial changes in the world economy and society. Global technological breakthroughs are causing substantial shifts in external and internal application models that promote better interactions in digital processes.

Financial technology (abbreviated Fintech) is one of the most widely used terms for research nowadays. Fintech refers to new technology that aims to enhance and streamline the delivery and usage of financial products and services. Fintech, at its heart, is used to assist corporations, company owners, and consumers in improving their financial affairs, procedures, and conduct. Fintech's prominence surged in the 2010s, and the global financial crisis of 2008 was the key catalyst for the evolution of Fintech (Haddad and Hornuf, 2019). The explanation for "why Fintech is happening today" is that Fintech groundbreaking concepts appear more authentic than earlier financial services advances (Rupeika-Apoga et al., 2018).

Between 2012 and 2021, the Fintech sector grew at an astonishing rate, including adoption rates, the number of Fintech businesses, and investments. Since 2017, the global Fintech industry's revenue has nearly doubled. In 2022, the worldwide Fintech sector's total revenue was roughly USD 194.1 billion and is predicted to rise at a CAGR of 16.8% between 2023 and 2028 and will reach USD 492.81 billion by 2028 (Figure 3.1).

Another recent development is Blockchain, which allows firms to create needed information for reporting straight from their financial records. Accounting, auditing, and finance have been recognized as industries that might benefit substantially from the distributed registry and other aspects of this specific technology as Blockchain technology proceeds to expand and be utilized in increasing numbers of corporate sectors. It is an ensemble of connected blocks on a network that connects peers to peers, wherein the blocks are linked via encrypted hash code, resulting in

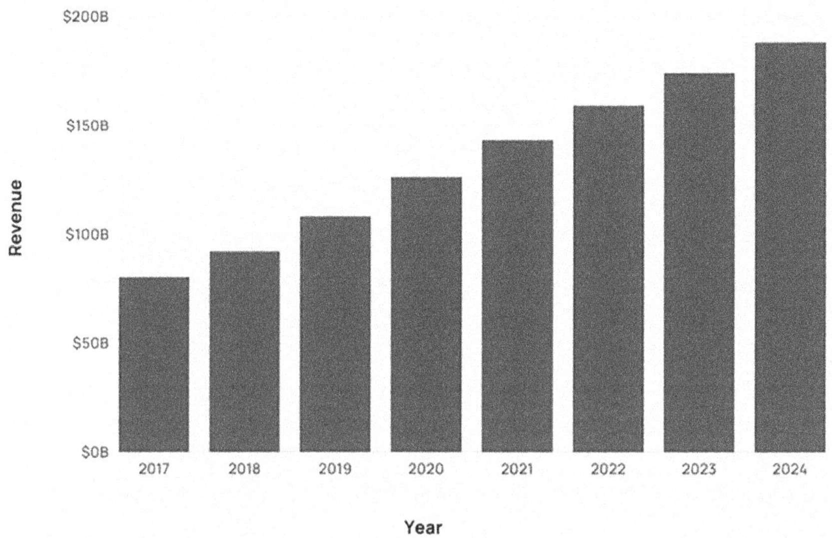

FIGURE 3.1 Fintech industry size.

Source: https://explodingtopics.com/blog/fintech-stats (2023).

no single point of failure. Because regardless of whether certain nodes in the network of peer-to-peer fail, the remaining nodes will continue functioning, which makes Blockchain more credible to manage and operate (Nakamoto, 2009). The Blockchain's second important aspect is that it is a digital technology that may be used for a number of purposes. Another important aspect is transparency and immutability; transactions made on the Blockchain are verifiable and accessible to those who use it, and it is nearly impossible to modify the transactions or the Blockchain itself (Skan, 2014).

Several studies have addressed the trends and issues of Fintech and Blockchain. Let's discuss a few of them briefly. Skan (2014) provide evidence that Fintech was growing at a rate that was more than double that of the rest of Europe. It has risen at double the pace of Silicon Valley since 2008. Since 2011, the amount of Fintech agreements in London has increased, accounting for more than half of total activities in Europe. Fintech firms have provided traditional banks an opportunity to move digitally and deliver affordable financial products and services. Low-cost banking with Fintech will enable banks to cope with more well-known and reputable banks in a nation like the Czech Republic (Hes and Jilkova, 2016).

There is an opportunity for emerging financial enterprises, which may be realized by means of a diverse variety of services and client appreciation. It is essential for the continuous growth of banks and financial services firms because the growth and advancement of Fintech firms will present clients with the choice to access both conventional financial institutions and revolutionary and imaginative products and services supplied by Fintech firms. Ramachandran (2019) analyzed the following five innovations as significant facilitators for potential Fintech in the setting of the UK: Blockchain, Unmanned Aerial Vehicle (UAV), Internet of Things (IoT), Robots, and Artificial intelligence (AI). Lee et al. (2021) studies the technology used in China's banking industry and how the expansion of the Fintech business impacts the efficiency of costs. According to the findings, commercial banking institutes owned by the state employ substandard technologies and have the lowest cost effectiveness. Stankevičienė and Kabulova (2022) drew the conclusion that when market factors have not been taken into account, there will not be any impact on financial institution potential due to unforeseen disruptions in Fintech advancement. Secondly, in developed nations, Fintech development may mitigate (or raise) the risk tolerance of financial markets, and the earnings potential of Fintech influences the resilience of financial institutions. Using a panel of 78 countries, Hodula (2022) indicates that credit platforms provided by Fintech businesses may serve as alternatives as well as complements to traditional bank loans, and that aspects of the banking industry are probably going to have a role in determining the link between them. Mirza et al. (2023) investigated the association between financial technology adoption and profitability for banks utilizing a panel fixed effects regression approach. The findings indicate an advantageous link between financial technology investment and ecological lending, which can be associated to the scrutiny, search, and tracking of the effectiveness regarding innovative technologies. According to Sharma et al. (2023), Green funds and Artificial intelligence are intended for long-term diversification investments, whilst Fintech is ideal for investing in both short- and long-term securities.

Blockchain technology is a hot topic for researchers and innovators. Evans (2015) evaluates a Blockchain management system (BMS) for accounting. On an everyday basis, new opportunities for Blockchain use in the financial industry become apparent. One example of this occurs in the field of crowdfunding; Muneeza et al. (2018) conducted a study on the use of Blockchain in crowdfunding. Utilizing data from six crowdfunding sites, research reveals that the adoption of Blockchain technology in crowdfunding can reduce the challenges encountered. In a similar vein, Alidin et al. (2018) suggested that linking Blockchain into conventional financial systems can be a difficult undertaking. Initiatives have been undertaken to link preexisting digital solutions with Blockchain using software developed by third parties. Laroiya et al. (2020) investigate several business domains wherein Blockchain technology (BCT) is useful, with a focus on the advantages that this technology gives to a number of corporations, companies, and society in general, as well as how such technology could enhance the way people live. Javaid et al. (2022) discusses Blockchain technology and its significance in the field of financial services. In addition, it examines different resources, techniques, and included services related to financial services based on Blockchain, as well as analyzes and assesses major uses of Blockchain in financial services. Ren et al. (2023) shows Blockchain technology has already been extensively utilized in a variety of businesses concerned with sustainable financing and has long-term implications for smart cities along with collaborative economies.

Considering the potential of such technology, there has been limited academic research and publications in this sector. The terms “Blockchain” and “Fintech” together have just 499 documents in the Web of Science database (July 27, 2023). Furthermore, because just 2.3% of these publications focus on business, management, and accounting, there is definitely an intense requirement for more investigation in this area of study. From theoretical and practical perspectives, deeper investigation into Fintech and Blockchain technology is required. We aim to accomplish this specifically by addressing the following five questions:

1. **RQ1:** What is the current pattern of Fintech and Blockchain technology research publication?
2. **RQ2:** Which countries and institutions have contributed most to Fintech and Blockchain technology research?
3. **RQ3:** What is the trend of publication in the most prevalent journals on Fintech and Blockchain technology research?
4. **RQ4:** Who are the most prominent authors on Fintech and Blockchain technology research?
5. **RQ5:** What are the trend topics on Fintech and Blockchain technology research?

In the following ways, this research varies from the earlier studies. We commence with establishing a more systematic understanding of Fintech and Blockchain technology research by bibliometric analysis, identifying present and prospective tendencies from 2016–2023. In addition, tables and diagrams have been created to offer an in-depth comprehension of the discipline's evolution as well as to identify research gaps in the present literature in order to suggest research pathways for subsequent investigations.

The rest of this work is structured as follows: the second section describes the research methods and findings; the third section focuses on the findings related to Fintech and Blockchain technology research, and the final section summarizes findings.

3.2 METHODOLOGY

3.2.1 LITERATURE RETRIEVAL PROCESS

Figure 3.2 depicts the research's approach of literature retrieval and bibliometric evaluation. We particularly retrieved key data on Fintech and Blockchain research technology articles from the Web of Science (WoS). The search string that was used for the first search was TITLE-ABS-KEY= (ALL=("Fintech") AND ALL=("Blockchain"), Time Span = 2016–2023, Document Type = (journal articles, book chapters, and proceeding papers). Indexes SCI-EXPANDED, SSCI, A&HCI, BKCI-S, BKCI-SSH, ESCI. A total of 499 articles were acquired; they were evaluated to provide 453 articles that were retained to comply with the study's objectives.

The basic information of sample articles is shown in Table 3.1. The subsequent part will discuss the bibliometric approaches utilized, namely VOSviewer and R Bibliometric.

3.2.2 BIBLIOMETRIC TECHNIQUE

Bibliometric analysis is a frequent and thorough method for studying and evaluating large volumes of data and producing high research impact. It makes it possible for academics to identify current developments in that field, authors' and countries' collaboration patterns, performance of journals and research elements, as well as investigate

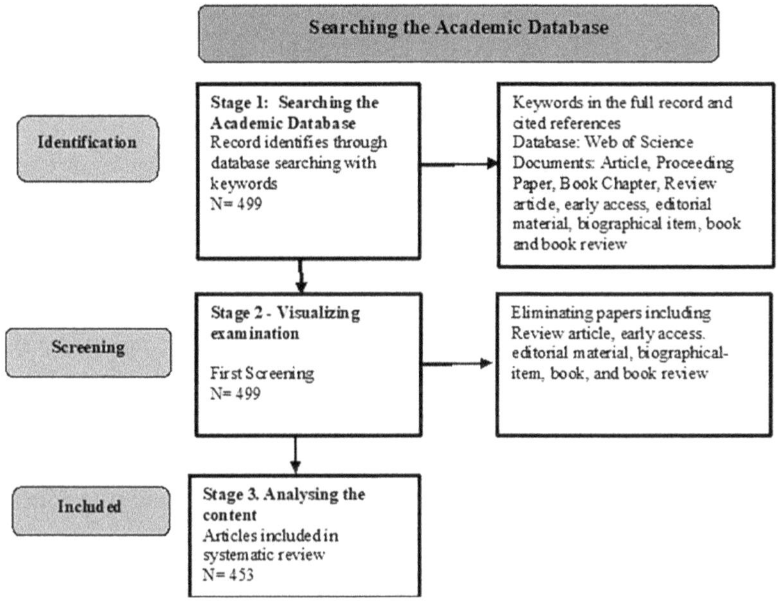

FIGURE 3.2 PRISMA flowchart for bibliometric analysis of Fintech and blockchain research technology.

Source: Author's compilation.

TABLE 3.1 Basic Information About Data, Documents, Authors, and Author's Collaboration

Description	Results
Time span	2016–2023
Sources	289
Documents	453
Annual Growth Rate %	65.49
Document Average Age	2.68
Average Citation Per Document	13.45
References	19671
Authors Keyword	1289
Keyword Plus	642
Authors	1184
Single-Authored Documents	84
Co-Authors Per Documents	3.07
International Co-Authorship %	35.1

Source: Author's compilation.

the conceptual framework of a certain area in the existing literature (Donthu et al., 2021; Verma and Gustafsson, 2020). The VOSviewer and Bibliometric R software tools are used to answer the research questions such as:

1. **VOSviewer:** Nees Jan van Eck and Ludo Waltman from Leiden University built VOSviewer in 2010 van Eck and Waltman (2010). This application allows users to explore and create maps utilizing data from networks. Although it is designed especially for examining academic records, it can additionally be employed to analyze any sort of network data (e.g., social networks). VOSviewer assesses citations, co-citations, co-occurrence, bibliographic coupling, and co-authorship links in the network, overlay, and density visualization modes.
2. **Bibliometric R:** Massimo Aria and Corrado Cuccurullo founded and created the bibliometrix R-package, that offers a variety of tools for quantitative study in bibliometrics and scientometrics. R, a programming language, was used to create the package, which is a freely available platform. The availability of significant, powerful statistical algorithms, availability of superior arithmetic procedures, as well as integrated data visualization capabilities are possibly the most compelling reasons to choose R over other programming languages for data analysis.

3.3 FINDINGS

3.3.1 ANNUAL SCIENTIFIC PRODUCTION ON FINTECH AND BLOCKCHAIN RESEARCH TECHNOLOGY

As can be observed in Figure 3.3, researchers from all domains are becoming increasingly interested in Fintech and Blockchain research technology. COVID-19 has dramatically transformed the Fintech and Blockchain industries. Customers enrolling and using smartphones for payments and app-based banking reached an all-time high in 2020, according to financial organizations. To prevent the spreading of COVID-19, governments all around the world are encouraging contactless payments. As a result, there will be a significant rise in publishing after 2020.

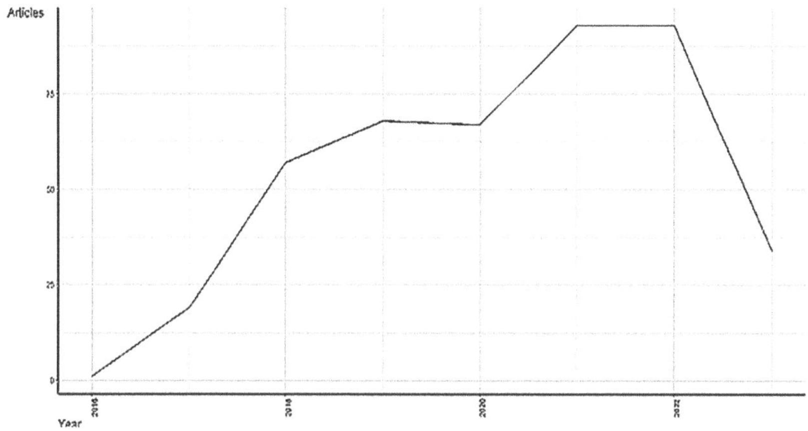

FIGURE 3.3 Annual number of academic publications on Fintech and blockchain.

Source: Author's compilation.

3.3.2 TOP INFLUENTIAL COUNTRIES AND INSTITUTIONS IN FINTECH AND BLOCKCHAIN RESEARCH TECHNOLOGY

3.3.2.1 TOP INFLUENTIAL COUNTRIES

The rapid expansion of Fintech and Blockchain literature in recent years is noteworthy despite the fact that they are still relatively new. Therefore, it is crucial to identify the countries and institutions that are home to the top Fintech and Blockchain research centers. The top 10 countries conducting Fintech and Blockchain research are listed in Table 3.2. Country (A) shows the top countries in terms of publications, whereas country (B) shows the top countries in terms of citations. *China* and *USA* are top nations in terms of both number of publications and citations, although the *UK* lags behind both countries in number of citations.

TABLE 3.2 Top 10 Countries Conducting Fintech and Blockchain Research

Rank	Country (A)	PFT*	Country (B)	TLC*
1	China	115	China	1354
2	USA	45	USA	1351
3	UK	26	France	494
4	Singapore	13	UK	376
5	Australia	13	Italy	331
6	Germany	11	Australia	258

TABLE 3.2 *(Continued)*

Rank	Country (A)	PFT*	Country (B)	TLC*
7	India	24	Spain	227
8	Vietnam	5	Turkey	213
9	Korea	10	Singapore	157
10	Russia	20	Canada	139

Source: Author's compilation.

Figure 3.4 depicts the collaborative world map, which represents the authors' country-based association. The deeper tone conveys more intense association, while the bright tone conveys weak association, and the grey color conveys absence of connection. In terms of country's collaboration, *"China"* and *"UK"* have the strongest links. *China*, being one of the most engaged nations in Fintech and Blockchain, has connections with *Pakistan (18), USA (16), Singapore (13),* and *Australia (11),* respectively. Three unique key cooperations (with America, Asia, and Australia) have been formed. Considering that every nation speaks different languages, Chinese intellectuals are more inclined to form global alliances.

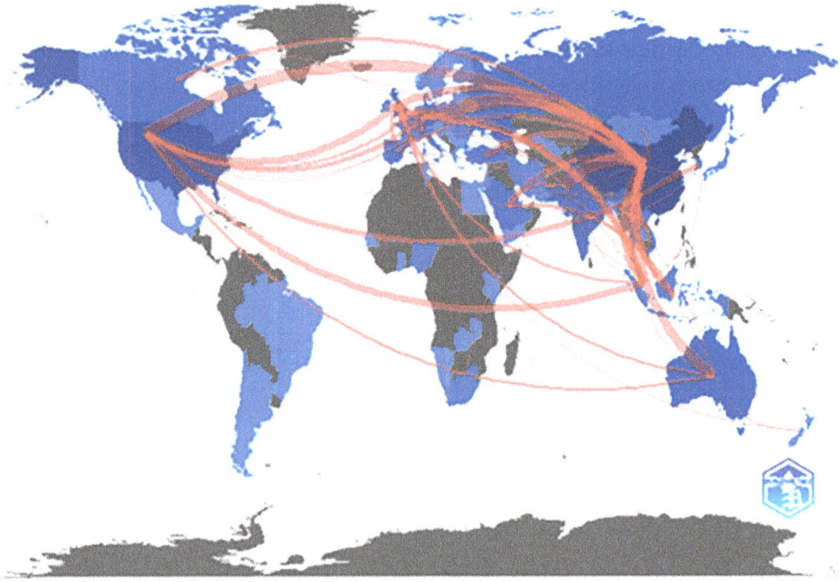

FIGURE 3.4 Collaboration network.

Source: Author's compilation.

3.3.2.2 TOP INSTITUTIONS

Table 3.3 shows the top 10 institutions. In the list, institution (A) represents the top 10 institutions with the most publications, whereas institution (B) represents the institutions with the most citations. *Asia University, Hang Seng University,* and *China Medical University Hospital* rank top in terms of both publications and citations. It is interesting to note that Fintech and Blockchain are being investigated not just by banking, financial, and management institutions, but also by medical, technological, post, and telecommunications universities.

TABLE 3.3 The Top 10 Institutions

Ranks	Institution (A)	PFT	Institution (B)	TLC
1	Asia University	114	Asia University	53
2	Hang Seng University	47	Hang Seng University	47
3	China Medical University Hospital	38	China Medical University Hospital	38
4	University of Hong Kong	14	University of Hong Kong	12
5	Nanyang Technological University	13	Singapore Management University	11
6	Singapore Management University	11	China Medical University	10
7	China Medical University	10	Complutense University of Madrid	10
8	Humboldt University	10	Peking University	8
9	Peking University	10	Humboldt University	7
10	Cornell University	9	Nanjing University of Post and Telecommunication	7

Source: Author's compilation.

3.3.3 MOST PRODUCTIVE JOURNALS IN FINTECH AND BLOCKCHAIN TECHNOLOGY RESEARCH

The more articles and citations a journal publishes, the more productive and effective it becomes. As a result, the number of publications, citations, and indexes connected with journal citations are all examined. Table 3.4 lists the most influential journals in Fintech and Blockchain Technology Research. Upon comparing the journals, *the Handbook of Blockchain, Digital Finance, and Inclusion* ranked first with 21 publications.

Sustainability and *Journal of Risk and Financial Management* are other relevant institutions that published articles related to Fintech and Blockchain technology research with 14 and 12 publications, respectively. *Review of Financial Studies* has six publications but still has the highest number of citations (772).

TABLE 3.4 The Top 10 Productive Journals

Rank	Name of Journal	Number of Publications	Number of Citations
1	Handbook of Blockchain, Digital Finance, and Inclusion	21	137
2	Sustainability	14	141
3	Journal of Risk and Financial Management	12	57
4	Annals of Financial Economics	9	11
5	Electronic Commerce Research and Applications	6	174
6	Review of Financial Studies	6	772
7	Technological Forecasting and Social Change	6	346
8	Financial and Credit Activity-Problems of Theory and Practice	5	8
9	Fintech: Law and Regulation	5	1
10	International Journal of Environmental Research and Public Health	5	139

Source: Author's compilation.

3.3.4 MOST INFLUENTIAL AUTHORS ON FINTECH AND BLOCKCHAIN TECHNOLOGY RESEARCH

In terms of most influential authors, Table 3.5 illustrates the top authors, their affiliations, published works, local citations, and global citations. Amongst the authors who were examined in the study, "Lin William Cong" from the University of Chicago is the most prolific, with 49 papers published, local citation (26), and global citation (295). Followed by "Peter Gomber" with five publications and a total of 364 citations. The results reveal that the top 10 authors have made substantial contributions to the advancement of Fintech and Blockchain research technologies during the previous seven years.

TABLE 3.5 The Top 10 Authors

Rank	Authors	Year	No of Publications	Name of University/ Institution	Local Citations	Global Citations
1	Lin William Cong	2019	49	University of Chicago, Chicago	26	295
2	Peter Gomber	2018	5	Goethe University of Frankfurt, Germany	21	343
3	Saman Adhami	2018	5	Bocconi University, Milan, Italy	18	180
4	Mark A Chen	2019	5	Georgia State University, Atlanta	11	141
5	Bruno Biais	2019	5	Université Toulouse Capitole, TSM-R and HEC Paris	10	115
6	Ittay Eyal	2017	5	Department of Electrical Engineering at The Technion	9	145
7	Jonathan Chiu	2019	5	Bank of Canada	8	72
8	Anjan V. Thakor	2020	4	Washington University, United States	8	164
9	Victor Chang	2020	4	Teesside University, UK	8	124
10	Wenyu (Derek) Du	2019	4	Beihang University, China	6	103

Source: Author's compilation.

3.3.5 MOST TRENDING TOPIC ON FINTECH AND BLOCKCHAIN TECHNOLOGY RESEARCH

Figure 3.5 depicts the evolution of research publications over time in terms of keywords. Words like entrepreneurship, strategy, performance, technology, systems, management, and competitions indicate a growth in researcher interest in Fintech and Blockchain Technology Research. Terms like impact, competition, Blockchain, bitcoin, and innovation indicate a significant rise within the research subject, although it remains less than 50 terms within the time period studied. Other terms that appear often in the field but are on a downward trend include performance, systems, competition, and strategy.

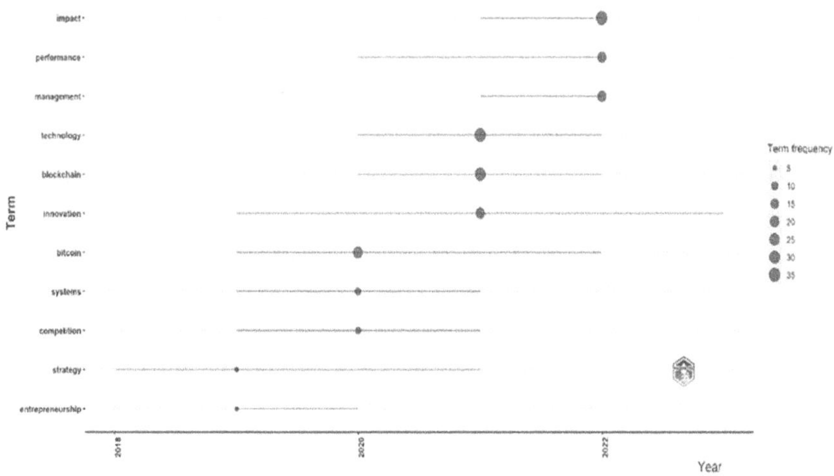

FIGURE 3.5 Keywords trend topic.

Source: Author's compilation.

3.4 CONCLUSION

According to the study, Fintech and Blockchain are the future of economics, finance, as well as information technology, and play a role in the safeguarding of our global ecosystem. The study identifies some significant and conceptual elements of the Fintech and Blockchain literature during the past seven years. We uncovered some of the key elements of Fintech and Blockchain literature. It is noteworthy that the expansion of Fintech and Blockchain is interesting, and research has risen significantly in the past few years.

We gave an overall evaluation of Fintech and Blockchain articles between 2016 and 2023 in this research. Based on WoS, we retrieved 453 papers that were relevant to the study's needs. First, the advancement of Fintech and Blockchain is aided by advances in several areas, including machine learning, artificial intelligence, big data, and the internet of things. Secondly, this field's study hotspots are evident and reflected from several perspectives: (i) from the top influential countries/regions (*see* Table 3.2), China and USA rank first in terms of both the number of publications and citations. Their studies focus on peer-to-peer lending, Blockchain, sustainability, and financial inclusion; (ii) from top institutions (*see* Table 3.3), Asia University made important contributions to research in this field, which is followed by Hang Seng University and China Medical University Hospital; (iii) from top productive Journals (*see* Table 3.4), the Handbook of Blockchain, Digital Finance, and Inclusion, Sustainability, and Journal of Risk and Financial Management are among the top prolific publications in this field, focused on Blockchain, online commerce, Fintech, the internet of things, banking, and large-scale data; and (iv) from top authors (*see* Table 3.5), the most prolific author is Lin William Cong from the University of Chicago. Top authors in the field also include Peter Gomber of Germany and Saman Adhami of Italy. Moreover, technology, impact, and Blockchain are the trend topic keywords but having frequency less than 50.

According to the study practicality, recent data on different academic institutions and researchers could prove valuable for Fintech and Blockchain experts and prospective businesses for future alliances, upgrading, and evaluation of ideas throughout numerous phases of business. This knowledge on the field may be valuable for new academic scholars in initiating future study effort. In addition, this type of data may be beneficial to potential opportunities for networking among scholars and other stakeholders.

3.5 FUTURE RESEARCH AGENDA

As with every study, there are limitations, and this research study was no exception. Firstly, to find studies that were appropriate for our study, we simply utilized WoS as a repository. Fintech and Blockchain are rapidly developing fields, and data is being ignored in other repositories such as Google Scholar, Scopus, and JSTOR. As a result, any future research might aggregate data from all of these databases to undertake a more thorough analysis. Secondly, the query being searched was carried out on 10 May 2023; this is a drawback because studies could change when new publications are uploaded. Lastly, future studies can be carried out to identify proliferation as well as bottlenecks of Industry 4.0 using the internet of things, Blockchain, and technological advancement mechanisms and E-commerce proliferation: filling gaps in government laws concerning payment via the internet, Blockchain and bitcoin permeability, identification of fraud, security measures, and restructuring financial orders.

KEYWORDS

- **bibliometric analysis**
- **blockchain**
- **FinTech**
- **keywords**
- **Rstudio**
- **VOSviewer**

REFERENCES

Alidin, A., Al-Wosabi, A., & Yusoff, Z. (2018). Overview of blockchain implementation on Islamic finance: Saadiqin experience. https://doi.org/10.1109/CR.2018.8626822 (accessed on 26th July 2025).

Bai, C., Dallasega, P., Orzes, G., & Sarkis, J. (2020). Industry 4.0 technologies assessment: A sustainability perspective. *International Journal of Production Economics*, *229*, 107776. https://doi.org/10.1016/j.ijpe.2020.107776 (accessed on 26th July 2025).

Brettel, M., Friederichsen, N., Keller, M., & Rosenberg, M. (2014). How virtualization, decentralization and network building change the manufacturing landscape. *International Journal of Mechanical, Aerospace, Industrial and Mechatronics Engineering*, *8*(1), 37–44.

Donthu, N., Kumar, S., Pandey, N., & Lim, W. M. (2021). Research constituents, intellectual structure, and collaboration patterns in *Journal of International Marketing*: An analytical retrospective. *Journal of International Marketing*, *29*(2), 1–25. https://doi.org/10.1177/1069031X211004234 (accessed on 26th July 2025).

Evans, C. W. (2015). Bitcoin in Islamic banking and finance. *Journal of Islamic Banking and Finance*, *3*(1), 1–11. https://doi.org/10.15640/jibf.v3n1a1 (accessed on 26th July 2025).

Haddad, C., & Hornuf, L. (2019). The emergence of the global fintech market: Economic and technological determinants. *Small Business Economics*, *53*(1), 81–105. https://doi.org/10.1007/s11187-018-9991-x (accessed on 26th July 2025).

Hodula, M. (2022). Does fintech credit substitute for traditional credit? Evidence from 78 countries. *Finance Research Letters*, *46*, 102469. https://doi.org/10.1016/j.frl.2021.102469 (accessed on 26th July 2025).

https://explodingtopics.com/blog/fintech-stats. (2023). 57+ incredible fintech stats (2024–2026). *ExplodingTopics*. https://explodingtopics.com/ (accessed on 26th July 2025).

Javaid, M., Haleem, A., Singh, R. P., Suman, R., & Khan, S. (2022). A review of blockchain technology applications for financial services. *BenchCouncil Transactions on Benchmarks, Standards and Evaluations*, *2*(3), 100073. https://doi.org/10.1016/j.tbench.2022.100073 (accessed on 26th July 2025).

Koesworo, Y., Muljani, N., & Ellitan, L. (2019). Fintech in the industrial revolution era 4.0. *International Journal of Research Culture Society*, *3*(9), 53–56. https://amp.kompas.com/tekno/read/2018/12/06/11560017/apa-kelebihan-dana- (accessed on 26th July 2025).

Laroiya, C., Saxena, D., & Komalavalli, C. (2020). Chapter 9–Applications of blockchain technology (S. Krishnan, V. E. Balas, E. G. Julie, Y. H. Robinson, S. Balaji, & R. B. Kumar, Eds.; pp. 213–243). *Handbook of Research on Blockchain Technology*. Academic Press. https://doi.org/10.1016/B978-0-12-819816-2.00009-5 (accessed on 26th July 2025).

Lee, C. C., Li, X., Yu, C. H., & Zhao, J. (2021). Does fintech innovation improve bank efficiency? Evidence from China's banking industry. *International Review of Economics & Finance*, *74*, 468–483. https://doi.org/10.1016/j.iref.2021.03.009 (accessed on 26th July 2025).

Mirza, N., Umar, M., Afzal, A., & Firdousi, S. F. (2023). The role of fintech in promoting green finance, and profitability: Evidence from the banking sector in the Euro zone. *Economic Analysis and Policy*, *78*, 33–40. https://doi.org/10.1016/j.eap.2023.02.001 (accessed on 26th July 2025).

Muneeza, A., Arshad, N. A., & Arifin, A. T. (2018). The application of blockchain technology in crowdfunding: Towards financial inclusion via technology. *International Journal of Management and Applied Research*, *5*(2), 82–98. https://doi.org/10.18646/2056.52.18-007 (accessed on 26th July 2025).

Nakamoto, S. (2009). Bitcoin: A peer-to-peer electronic cash system. *Cryptography Mailing List at https://metzdowd.com*. https://bitcoin.org/bitcoin.pdf (accessed on 26th July 2025).

Ramachandran, M. (2019). Reference architecture and business process modelling for fintech applications. *July*. http://eprints.leedsbeckett.ac.uk/id/eprint/6390/ (accessed on 26th July 2025).

Ren, Y. S., Ma, C. Q., Chen, X. Q., Lei, Y. T., & Wang, Y. R. (2023). Sustainable finance and blockchain: A systematic review and research agenda. *Research in International Business and Finance*, *64*, 101871. https://doi.org/10.1016/j.ribaf.2022.101871 (accessed on 26th July 2025).

Rupeika-Apoga, R., Zaidi, S. H., Thalassinos, Y. E., & Thalassinos, E. I. (2018). Bank stability: The case of Nordic and non-Nordic banks in Latvia. *International Journal of Economics and Business Administration*, *6*(2), 39–55. https://doi.org/10.35808/ijeba/156 (accessed on 26th July 2025).

Saucedo-Martínez, J. A., Pérez-Lara, M., Marmolejo-Saucedo, J. A., Salais-Fierro, T. E., & Vasant, P. (2018). Industry 4.0 framework for management and operations: A review. *Journal of Ambient Intelligence and Humanized Computing*, *9*(3), 789–801. https://doi.org/10.1007/s12652-017-0533-1 (accessed on 26th July 2025).

Sharma, S., Aggarwal, V., Dixit, N., & Yadav, M. (2023). Time and frequency connectedness among emerging markets and QGREEN, FinTech and artificial intelligence-based index: Lessons from the outbreak of COVID-19. *Vision: The Journal of Business Perspective*. https://doi.org/10.1177/09722629221141553 (accessed on 26th July 2025).

Stankevičienė, J., & Kabulova, J. (2022). Financial technology impact on stability of financial institutions. *Technological and Economic Development of Economy*, *28*(4), 1089–1114. https://doi.org/10.3846/tede.2022.17093 (accessed on 26th July 2025).

van Eck, N. J., & Waltman, L. (2010). Software survey: VOSviewer, a computer program for bibliometric mapping. *Scientometrics*, *84*(2), 523–538. https://doi.org/10.1007/s11192-009-0146-3 (accessed on 26th July 2025).

Verma, S., & Gustafsson, A. (2020). Investigating the emerging COVID-19 research trends in the field of business and management: A bibliometric analysis approach. *Journal of Business Research*, *118*, 253–261. https://doi.org/10.1016/j.jbusres.2020.06.057 (accessed on 26th July 2025).

CHAPTER 4

Financial Education Powered by Blockchain Technology

CHANCHAL PHORE,* PALLAVI RAJAIN, and PRIYANKA

Maharaja Surajmal Institute, New Delhi, Delhi, India

*Corresponding author

ABSTRACT

Blockchain technology has the potential to revolutionize the way one learns about finance. By providing a secure, transparent, and tamper-proof platform for storing and sharing financial data, blockchain can make financial education more accessible, engaging, and effective. One of the most significant benefits of blockchain for financial education is that it can make learning more accessible. In the past, financial education has been limited to those who could afford to pay for it. for storing and sharing financial data, blockchain can make financial education more accessible, engaging, and effective. The main aim of this study is to find out how to create more accessible financial education materials Blockchain-based financial education platforms provide people with access to high-quality financial education resources. Secondly, to analyze how to make financial education more engaging. Blockchain can create more engaging financial education experiences by allowing users to interact with financial data innovatively Overall, blockchain technology can potentially revolutionize how we learn about finance. Blockchain-based financial institutions are providing people with access to financial services that are more transparent and efficient. By providing a secure, transparent, and tamper-proof platform for storing and sharing financial data, blockchain can make financial education more accessible, engaging, and effective.

FinTech and Blockchain Technology in Financial Markets. Neha Puri, Anjali Munde, Emilia Alaverdov (Eds.)
© 2027 Apple Academic Press, Inc. Co-published with CRC Press (Taylor & Francis)

DOI: 10.1201/9781779641298-4

4.1 INTRODUCTION

Financial literacy refers to people's knowledge and understanding of basic financial terms and concepts (Hastings et al., 2013). Financial literacy is one of the major concerns, even for well-developed nations, as it affects not only the personal life of the people but also society at large. The lower levels of financial literacy among people lead to ineffective spending and financial planning, as well as expensive borrowing and debt management. Financial literacy will lead to financial inclusion, which is important as financial inclusion provides access, whereas financial literacy provides awareness (Subbarao, 2013).

Immediate efforts are needed to tackle the issue of financial illiteracy among people, and for this, more financial awareness and programs need to be spread. Financial education should be delivered to the students in schools at a young age, and blockchain technology can be used effectively. Blockchain is continuously implemented in various applications, such as healthcare, banking, and the Internet. Blockchain can be defined as a public ledger or distributed database wherein all the information is recorded in a chain of blocks and can be accessed by all the users in the network. This innovative technology functions decentralized, and the transactions entered are kept anonymous and immutable (Yaga et al., 2019). Since the transactions cannot be altered, the users need to enter new transactions to fix any error, hence trust in its users. The information accessed by the different parties is of the same level.

Every field is now technologically driven, so why not education? Education is one the most important sectors of any economy, and efforts should be made to deliver it to the students in the best possible way. The features like transparency, accessibility, security, and so forth, make blockchain technology a perfect fit for education. This chapter focuses on the effective applications of Blockchain in financial education. Financial education involves the skills, knowledge, and attitude necessary for understanding and managing one's money. Financial education plays an important role as it prepares people, mostly students, to face the wider world by helping them make sound financial decisions. The knowledge gained and skills acquired under financial education will lead to financial literacy, as financial literacy is a process of acquiring financial concepts, knowledge, and skills. With the help of basic personal finance skills, including spending, investing, saving, or budgeting, the young generation can easily make informed decisions and cope with financial issues. They can manage their debts and save for their future.

Blockchain is a much more effective and efficient technique than traditional databases as in traditional databases; the information could be tampered with or edited because it uses a client-server network. If we talk about Blockchain, then the users cannot edit the information stored in blocks, and for any editing, the user must store new transactions with the right information. The information is decentralized in blockchain technology, giving each user access to the same information. Anyone who tries to change the data is automatically corrected whenever inconsistent information is detected. Due to these reasons, users trust this inventive technique easily and adapt it worldwide in different sectors. Also, the information stored in blocks is more organized and paperless than its counterparts.

This chapter focuses on how Blockchain can be effectively used to create an accessible and secure environment in the field of financial education. This chapter is structured as follows. The first section of this chapter discusses the problem of financial illiteracy among nations and provides the solution for the same by providing financial education through blockchain technology. In the next section, the authors have reviewed the existing literature on Blockchain in education and Blockchain and financial literacy. The next section sheds light on the current scenario and applications of Blockchain in financial education, along with relevant examples. The chapter then explains the challenges in this field, and finally, the authors have given conclusory remarks.

4.2 LITERATURE REVIEW

4.2.1 BLOCKCHAIN IN EDUCATION

Ramos and Queiroz (2022) investigated the influence of trust on the adoption and implementation of Blockchain in education. An exploratory qualitative approach was used wherein educators of higher educational institutions in Brazil were questioned and interviewed. It was found that trust was an influential factor motivating the use of blockchain technology. The study suggested that the use of Blockchain would increase social relationships. Previously, Alammary et al. (2019) systematically reviewed the literature to investigate blockchain-based educational applications.

Three main themes were the main focus of the study: the development of educational applications using blockchain technology, benefits delivered by blockchain technology in education, and challenges faced when using blockchain technology. A detailed discussion and intensive analysis were conducted for each of these themes. Along similar lines, Bhaskar et al. (2020) provided a detailed systematic literature review on Blockchain in education. The authors also conducted a bibliometric analysis using the Scopus database on "Blockchain technology in Education." The study identified barriers and benefits along with the present application of Blockchain in education.

The research in this area has picked up in recent times, with several researchers focusing on the applications of Blockchain. Ocheja et al. (2022) used bibliometric and qualitative analysis of Blockchain in education. The authors identified major actors in the space, demographic participation and adoption, and potential areas of innovation. The results of the research revealed that most of the studies focused on reporting and verifying academic certificates. Several researches focused on the in-depth connection of academic records through assessment data, learning content, and learning behavior logs. Panagiotidis (2022) focused on the current and potential applications of Blockchain in education. This chapter analyzed the advantages and features of blockchain technology. Nowadays, blockchain applications are concerned with the administration of academic processes, payment of fees for studies, learning accreditation, cooperation between students and faculty members, protection of intellectual property, academic degree management, and evaluation of learning outcomes. Further, Ocheja et al. (2022) presented a unique qualitative and bibliometric analysis of Blockchain in education, as this topic has become popular only in the past five years. The research also revealed that most studies focused on reporting and verifying academic certificates and transcripts, with only a few researches dealing with assessment data, learning content, and behavior logs. Current education needs to consider interoperability at the blockchain level and the heterogeneous nature in which institutes create and consume data.

4.2.2 BLOCKCHAIN AND FINANCIAL LITERACY

Mhlanga (2023) analyzed how blockchain technology helped inclusion in the financial system of previously underserved populations. The study's findings showed that even though financial inclusion was not part of the global goals, they still included financial services for most of the population. Previously, Park (2021) explored the impact on education with the use of blockchain technology. Novel standards were set for crypto-learning and crypto-administration with its distributed education blockchain across organizations. The slow rate of blockchain adoption in technology is reflected in finance and management, which posed challenges like "blockchain mining" with less orientation towards collective development in education.

Earlier, Schuetz and Venkatesh (2019) studied rural India's economic development by connecting remote villages with global supply chains. They reviewed the literature on financial inclusion and the adoption of Blockchain in India. The focus was on four major challenges: financial illiteracy, inappropriate banking products, high cost, and geographic access. The answer to these questions ultimately leads to the adoption of Blockchain in rural areas. Recently, Weerawarna et al. (2023) explored blockchain technology in finance, specifically focusing on how blockchain architecture aids in gaining an advantage in the financial sector. The authors analyzed the content of 50 reports and articles through literature from the past decade. This study highlighted the various dimensions of blockchain technology, the current status of finance, its competitive advantages, and numerous challenges faced at the initial stages. For blockchain technology to become "next-generation networks," the authors identified three areas. William and Mary (2023) described the importance of Blockchain as a fintech application for improving financial literacy. This article discussed that with the growing importance of financial literacy, innovations were observed in financial technologies and services. With the prioritization of financial education, not just through financial inclusion but also with the help of fintech companies, people can be empowered to take control.

4.2.3 RESEARCH GAP

An assessment of previous literature showed that the studies in the area of Blockchain in education and the role of Blockchain in finance were more prevalent. There is scant research interlinking both these fields. As there is a scarcity of literature focusing on the application of Blockchain in financial education, there is a requirement to explore this particular area to gain more insights.

4.3 CURRENT SCENARIO AND APPLICATION

4.3.1 INNOVATIVE LEARNING LANDSCAPES

The teaching and learning process can be made more interesting with the help of Blockchain. Gamification is one of the promising applications of Blockchain wherein mundane financial lessons can be made interesting and engaging by giving the students meaningful tasks to be completed. Then, students are rewarded with digital tokens, badges, or progress bars, which instill a sense of achievement. These game-based elements would boost their self-esteem and help them to remember and retain the tips and tricks for saving, investing, and managing debt. All this will ultimately lead to enhancement in the learning process and betterment in learning outcomes.

4.3.2 SELF-PACED LEARNING

Financial concepts sometimes become hard to cope with, and students might take time to absorb such concepts. In this context, Blockchain can help students to learn and understand the content at their own pace as the data is stored securely in the blocks. The records stored in the blocks helps the students to choose the course as per their requirement.

Another important aspect of self-paced learning is that the students tend to join a course online and then leave it in between as they lose interest and motivation to complete it. However, blockchain technology uses rewards, providing students with a sense of achievement; hence, they make progress continuously and complete the course on time.

4.3.3 SECURITY AND ACCESSIBILITY

The accessibility and availability of educational resources are no longer a concern. Blockchain technology has made accessing the needed resources from different geographical areas and socio-economic statuses easier. Blockchain technology will also be useful in providing and delivering financial education programs or material in different languages. Educational resources are being stored safely because of highly secure encryption methods and distributed networks used in blockchain technology.

This innovative technology helps the students to access the resources easily and in a safe manner as access is restricted to those who are authorized to use them.

4.3.4 RECORD MANAGEMENT

Blockchain technology would help maintain records, including transcripts, degrees, credits earned, report cards, skills acquired and so forth, chronologically for every student lifelong. The immutable blockchain technology assures that these documents cannot be manipulated or tampered with, so they can be trusted easily, and universities can easily move towards paperless offices.

The attendance records can also be stored in blocks, which helps ascertain the regularity of students, and the same can be informed to the concerned stakeholders. The assignments can also be stored in these blocks. All this data can be easily maintained without fearing losing them, and these stored digital documents are well organized.

The library is the soul of every academic institution; therefore, it is working should be as smooth as possible. Record keeping in libraries is essential for this smooth functioning. The information regarding several books available and issued to readers must be constantly stored and updated (Zhang, 2019).

4.3.5 BUILDS TRUST AND TRANSPARENCY

Blockchain technology has made it convenient to store lifelong academic records, including degrees, certificates, and so forth, of oneself. All the credits gained by people in their academic journey can be stored in blocks. This information can easily be used and shared with the institution in which the student is seeking admission for further studies or even with prospective employers. This feature is useful from the viewpoint of the student or employee, and from the perspective of the employer and admission provider, as it becomes easier for them to judge the educational background along with other skills of the person, and the same could be verified. This imparts a sense of transparency and integrity to users.

4.3.6 RATIONAL FINANCIAL DECISION MAKING

Blockchain technology and AI-driven Robo Advisors can be used together to make an informed financial decision. The algorithms used by robo-advisors will help in seeking and further analyzing the market for tradable use. Then, this information can be used by Blockchain to trade in crypto-currencies, just like trading in stock options or futures.

Another innovative way of using Blockchain in spreading financial education is through crypto-friendly banks. These are the financial institutions that cater to the needs of cryptocurrency users by giving them tailored services. In simple terms, these banks manage the digital assets of their users. These banks can provide the required knowledge to the investors regarding the risks related to cryptocurrencies and empower them to manage their finances well such as:

1. **Increasing Engagement in Learning Through Blockchain:**
   a. **Block Chain University:** Several universities offer education in Blockchain worldwide. For instance, IIT Kanpur offers certificate courses in blockchain technology and applications, IIT Madras provides certification in software engineering for Blockchain, Portland State University (USA) offers courses in business blockchain, Harvard University (USA) offers courses in introduction to blockchain bitcoin and breakthrough innovation with blockchain technology, University of Oxford (UK) providing certification blockchain strategy, University Canada West (Canada) offering certification in blockchain entrepreneurship.

   b. **Coin-Scribble:** With the aid of coin-scribble, a cryptocurrency press release distribution service, crypto companies can swiftly and effortlessly acquire coverage on thousands of blockchain news sources. Companies such as Tron, immutablex, and coin-bound have utilized the service.

   c. **Blockcerts:** A blockchain-based certificate's development, issuance, watching, and verification can all be accomplished via Blockcerts, an open standard. It allows educational institutions to generate tamper-proof, secure, and verified digital credentials.

d. **ODEM:** It is a worldwide educational marketplace that securely validates and exchanges academic credentials utilizing blockchain technology. It efficiently and transparently connects students, teachers, and service providers.

e. **Signals:** Blockchain platform Signals, which operates on Ethereum, is working on a project allowing cryptocurrency traders to employ AI and machine learning as an advantage when trading. The Signals platform will merely require traders to choose their trading strategies; highly precise computer algorithms will handle large data analysis and decision-making.

f. **Robo Advisor Coin:** An AI-powered blockchain platform named Robo Advisor Coin, which is currently in its final development stages, was first released in 2014. Once launched, it will utilize AI and machine learning to forecast the cryptocurrency markets while using blockchain technology to validate these predictions' reliability, honesty, and precision. The Robo Advisor Coin algorithms have been rigorously tested against historical data and have a proven track record of success.

g. **Numerai:** Cryptocurrencies are additionally utilized by Numerai, an AI-driven hedge fund, to compensate the data scientists who assist in creating its algorithms. Based on the Numerai business model, programmers are given access to encrypted data to develop trading algorithms and, in turn, gain the platform's token.

h. **Bitcoin Games:** Online gaming experiences that use Bitcoin or other cryptocurrencies as essential elements, either as in-game currency or as prizes, are called "Bitcoin games." Slot machines and table games are available at Bitcoin casinos like BitStarz and CryptoSlots. Games like Satoshi Dice and Sportsbet.io cater to sports fans hoping to wager on different occurrences using Bitcoin. At the same time, Satoshi Dice enables players to gamble on the outcomes of virtual dice rolls for a more collaborative experience. Virtual worlds like Decentraland and The Sandbox offer settings where users can buy and create virtual land. In contrast, blockchain-based RPGs like Crypto-Kitties and Axie Infinity offer unique in-game assets that players can purchase, sell, and trade using cryptocurrencies. Furthermore, some Bitcoin mining video games like Roller-Coin imitate the cryptocurrency mining process and let players win Bitcoin as they play.

i. **IBM Blockchain for Education:** For educational institutions, IBM offers blockchain solutions. The technology they use can be utilized to streamline administrative procedures, verify credentials, and store records in a secure manner.

j. **Crypto-friendly Banks:** Several crypto-friendly banks are available worldwide, like Kraken Financial, which obtains a special purpose depository institution (SPDI) charter in Wyoming, USA, allowing it to operate crypto-friendly bank services to traditional and cryptocurrency clients. New York-based Signature Bank, Arab Bank Switzerland, Fidor Bank Germany, and Silver-Gate Bank based in California, USA, are examples of crypto-friendly banks that provide services related to cryptocurrency in many ways possible.

k. **Sovrin:** With the help of blockchain technology, Sovrin, a decentralized identity platform, offers individuals ownership and control over their digital identities. This can be used in educational contexts to securely maintain student identities and credentials.

l. **BitDegree:** With the help of BitDegree, users may obtain tokens for completing courses on an online education network based on the Blockchain. These tokens, referred to as BitDegree Tokens (BDG), can be redeemed for discounts or utilized for accessing premium content on the network.

2. **Challenges:**

a. **Climate Change:** The scale and intensity of mining depend on the size of the blockchain; the greater the size of the blockchain, the greater the mining effort. Such mining efforts require 1,000 trillion solutions per second, leading to high electricity consumption. The high electricity consumption, in turn, leads to a greater carbon footprint, ultimately leading to climate change (Park, 2021).

b. **Authoritarian Social Control:** The greater aim of Blockchain should be to serve a sustainable education system with peer-to-peer cooperation. Instead, it sometimes falls prey to authoritarian social control and hyper-capitalism. The aim of blockchain adoption for financial literacy should not be just for technological novelty but for long-term education goals (Park, 2021).

c. **Inflexible Record:** The information, once entered in blocks, cannot be modified, which at that time may be a benefit, but in the long run, it poses a challenge. When there is a need to update the stored information, it may hamper the data's integrity because the modified transaction needs to be approved by all the members. If not done simultaneously, this approval may be considered illegitimate (Grech and Camilleri, 2017). Every University continuously revises its syllabus, removing some previous topics and adding new ones. Due to this revision, the consent of all users is not granted, resulting in the non-updation of new topics. The irrelevant topics in the present context still need to be clarified, leading to confusion.

d. **Scalability:** The educational institutes maintain the records of students, such as transcripts, degrees, certificates, fee-related information, and attendance records, using blockchain technology. The data for each student is huge, and every transaction is verified by each user, which leads to growth in the number of blocks. The speed of transactions is slowed down, which might challenge blockchain scalability (Steiu, 2020).

e. **Constant Innovation:** Some researchers have suggested that since Blockchain is a new technology, it is continually evolving and changing. Therefore, most of the time, projects based on blockchain technology fail or get scrapped. A careful risk analysis needs to be done before going ahead with any project based on Blockchain.

4.4 CONCLUSION

Technological advancements are taking place in every field and have impacted the way we work in our daily lives. The education field is no exception to this progress. Blockchain is one of the latest advancements which is now being used in education. Although this research is still nascent, the authors have elaborated on how blockchain technology can be useful in spreading financial education among the masses. The lack of knowledge regarding financial products available in the market brings significant costs to the nation and the individual. The authors have identified different areas where blockchain technology can be applied to financial education: Innovative Learning Landscapes, Self-Paced Learning, Security and Accessibility, Record Management, Builds Trust and Transparency, and Rational Financial Decision Making. All these applications will help to spread financial knowledge to the public at large at a minimized cost. As more and more people become financially literate, they can easily manage their investments, savings, and debt. They will also be aware of the complex financial products available and ultimately make an informed decision. Financial literacy will lead to financial inclusion, ultimately boosting the economy.

KEYWORDS

- **bitcoin**
- **blockchain**
- **decentralized**
- **financial education**
- **financial literacy**
- **fraud detection**
- **smart contracts**

REFERENCES

Alammary, A., Alhazmi, S., Almasri, M., & Gillani, S. (2019). Blockchain-based applications in education: A systematic review. *Applied Sciences*, *9*(12), 2400.

Bhaskar, P., Tiwari, C. K., & Joshi, A. (2021). Blockchain in education management: Present and future applications. *Interactive Technology and Smart Education*, *18*(1), 1–17.

Grech, A., & Camilleri, A. F. (2017). *Blockchain in Education*. Publications Office of the European Union: Luxembourg.

Hastings, J. S., Madrian, B. C., & Skimmyhorn, W. L. (2013). Financial literacy, financial education, and economic outcomes. *Annual Review of Economics*, *5*(1), 347–373.

IBM Blockchain Pulse. (2019). What's the difference between a blockchain and a database? https://www.ibm.com/blog/whats-the-difference-between-a-blockchain-and-a-database/ (accessed on 26th July 2025).

Ma, Y. (2023). The role of crypto-friendly banks in facilitating financial inclusion. https://www.plugandplaytechcenter.com/resources/crypto-friendly-banks/ (accessed on 26th July 2025).

Mhlanga, D. (2023). Blockchain technology for digital financial inclusion in the industry 4.0, towards sustainable development? *Frontiers in Blockchain*, *6*, 1035405.

Ocheja, P., Agbo, F. J., Oyelere, S. S., Flanagan, B., & Ogata, H. (2022). Blockchain in education: A systematic review and practical case studies. *IEEE Access*, *10*, 99525–99540.

Panagiotidis, P. (2022). Blockchain in education—The case of language learning. *European Journal of Education (EJE)*, *5*(1), 66–83.

Park, J. (2021). Promises and challenges of blockchain in education. *Smart Learning Environments*, *8*(1), 1–13.

Ramos, C. R. D. S., & Queiroz, M. M. (2022). Blockchain in education: The influence of trust on adoption and implementation. *RAUSP Management Journal*, *57*, 316–331.

Schuetz, S., & Venkatesh, V. (2020). Blockchain, adoption, and financial inclusion in India: Research opportunities. *International Journal of Information Management*, *52*, 101936.

Staff Writer. (2023). Best universities offering blockchain and crypto degrees. https://www.banklesstimes.com/cryptocurrency/resources/universities-offering-blockchain-and-crypto-degrees/ (accessed on 26th July 2025).

Steiu, M. F. (2020). Blockchain in education: Opportunities, applications, and challenges. *First Monday*. https://firstmonday.org/ojs/index.php/fm/article/view/10640 (accessed on 26th July 2025).

Tamplin, T. (2023). Blockchain in education. https://www.financestrategists.com/wealth-management/blockchain/blockchain-in-education/ (accessed on 26th July 2025).

Thayer, T. L. (2018). Promising and ambitious blockchain initiatives for higher education. *Gartner Research*, *20*.

Weerawarna, R., Miah, S. J., & Shao, X. (2023). Emerging advances of blockchain technology in finance: A content analysis. *Personal and Ubiquitous Computing*, 1–14.

William & Mary. (2023). Fintech: Shaping online financial education. https://online.mason.wm.edu/blog/fintech-shaping-online-financial-education (accessed on 26th July 2025).

Yaga, D., Mell, P., Roby, N., & Scarfone, K. (2019). Blockchain technology overview. *arXiv preprint*, arXiv:1906.11078. https://arxiv.org/abs/1906.11078 (accessed on 26th July 2025).

Zhang, L. (2019). Blockchain: The new technology and its applications for libraries. *Journal of Electronic Resources Librarianship*, *31*(4), 278–280.

Zheng, Z., Xie, S., Dai, H. N., Chen, X., & Wang, H. (2018). Blockchain challenges and opportunities: A survey. *International Journal of Web and Grid Services*, *14*(4), 352–375.

Zhu, W. (2023). Coinscribble review | Top blockchain & crypto PR distribution service? https://coinbound.io/coinscribble-review/ (accessed on 26th July 2025).

CHAPTER 5

Revolutionizing Fraud Prevention in Financial Systems Through Blockchain Techniques

CHABI GUPTA, FINCY PALLISSERY, and KRITI PARASHAR*

[1]*Amity College of Commerce and Finance, Amity University, Noida, India*

[2]*School of Law, CHRIST University Bangalore, India*

*Corresponding author

ABSTRACT

Blockchain has gained immense popularity recently. The use of blockchain techniques has gained popularity to prevent fraud and maintain transparency across various industries. These methods help combat fraudulent activities within the financial sector, where high-value transactions occur regularly. This research delves into the intricate workings of blockchain technology and its applications in financial systems, with a particular emphasis on managing information flow. The study examines five distinct levels of technological complexity to explore how blockchain implementations impact information management: from basic information storage through to advanced processing techniques that enhance data quality and lead towards seamless integration between financial systems. A blockchain has the potential to revolutionize the way financial services are provided by enabling transactions such as payments and securitization without relying on traditional banking institutions. It can eliminate fraud. Moreover, utilizing "smart contracts" based on the blockchain introduces automation. These self-executing contracts could help in compliance

FinTech and Blockchain Technology in Financial Markets. Neha Puri, Anjali Munde, Emilia Alaverdov (Eds.)
© 2027 Apple Academic Press, Inc. Co-published with CRC Press (Taylor & Francis)

DOI: 10.1201/9781779641298-5

management, claims processing, and distributing financial content from legal documents like wills. Instances of cybersecurity breaches continue to disrupt information flow globally. This trend highlights the need for greater depth in the approach to cybersecurity measures, recognizing that simply relying on one solution may not be sufficient in combating increasingly sophisticated threats.

5.1 INTRODUCTION

Blockchain is an innovative technology that has gained immense popularity in recent times, particularly within financial systems. This decentralized digital ledger offers a robust and secure method for recording transactions without the need for intermediaries. The use of blockchain techniques has emerged as a potent tool to prevent fraud and maintain transparency across various industries. These methods have proven invaluable in combating fraudulent activities within the financial sector, where high-value transactions occur regularly. Through its sophisticated algorithms and security measures, blockchain ensures that all transaction data remains tamper-proof while remaining accessible to authorized parties.

1. **Research Purpose/Objective:** This research delves into the intricate workings of blockchain technology and its applications in financial systems, with a particular emphasis on managing information flow. The study examines five distinct levels of technological complexity to explore how blockchain implementations impact information management: from basic information storage through to advanced processing techniques that enhance data quality and lead towards seamless integration between financial systems. Through rigorous analysis and innovative thinking, this research sheds new light on the potential for blockchain technologies to revolutionize the way fraud can be prevented in financial systems.
2. **Findings:** Blockchain has the potential to revolutionize the way financial services are provided by enabling transactions such as payments and securitization without relying on traditional banking institutions. This decentralized approach opens a world of possibilities, empowering financial systems with greater

control and eliminating incidents of fraud. Moreover, utilizing "smart contracts" based on the blockchain can introduce a level of automation that was previously unheard of. These self-executing contracts could potentially streamline various processes, including compliance management, claims processing, and even distributing financial content from legal documents like wills. By removing intermediaries and reducing human intervention, these smart contracts promise increased efficiency while minimizing errors or discrepancies.

3. **Recommendation:** Despite the rigorous implementation of blockchain technology, instances of cybersecurity breaches continue to plague financial systems and disrupt information flow on a global scale. This ongoing trend highlights the need for greater depth in the approach to cybersecurity measures, recognizing that simply relying on one solution may not be sufficient in combating increasingly sophisticated threats.

5.2 BACKGROUND

Blockchain sets itself apart from conventional databases by organizing and storing data in blocks, which are linked to one another when full. This unique chain structure system ensures maximum security and transparency for authorized network members accessing the data within (Abdelhalim and Ibrahim, 2023; Naik et al., 2020; Yamuna et al., 2023). The integrity of the recorded information remains immutable due to its incorruptible nature, making it a valuable tool for businesses looking to enhance their fraud detection and prevention capabilities, among other tasks. With these exceptional features at hand, companies can leverage this cutting-edge technology towards greater efficiency while maintaining unparalleled levels of trustworthiness in their operations.

Blockchain technology offers several benefits to companies, providing a secure and efficient means of managing data. Its ability to prevent hacks and fraud is one such advantage that stands out. By leveraging its advanced security measures, blockchain can offer protection against malicious attacks on company systems or information. Another benefit of using blockchain is the reduced need for centralized platforms or intermediary institutions, which leads to cost savings in the long run (Figure 5.1).

The technology's unique architecture restricts or blocks access by unauthorized users to sensitive corporate data. Additionally, transactions get faster thanks to blockchain technology's decentralized approach, as it eliminates intermediaries between parties involved in the transaction process. Furthermore, transparency plays a vital role since all transactions are transparently recorded on the ledger, making it easy for everyone involved to track every single aspect related to them (Liu et al., 2022; Sathya and Balakumar, 2022; Zelic and Baros, 2018). Given these advantages brought about by this transformative innovation and advancements made with fraud detection technologies, coupled with some real-world successful use cases, embracing Blockchain could be an effective measure towards detecting fraudulent activities within organizations rather than just traditional methods like audits alone (CEBECİ and ARAT, 2022; Charde and Hande, 2022; Hong Hin, 2020).

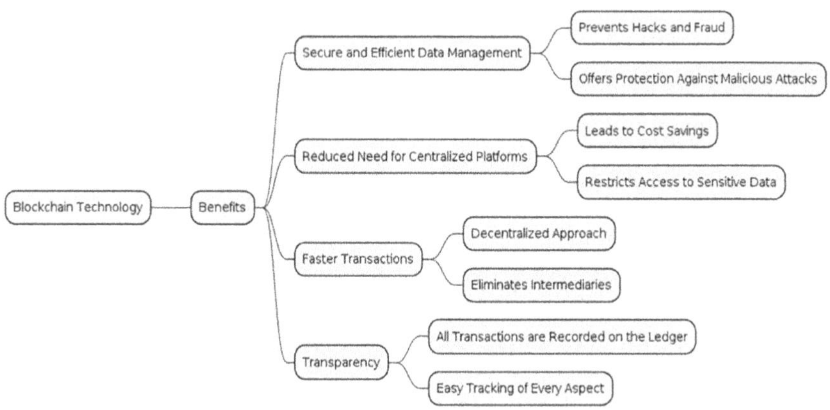

FIGURE 5.1 Benefits of using block chain in financial systems.

Source: Author's Research.

Distributed networks are a powerful tool for managing data and transactions. By allowing authorized members to monitor, share, and control information across multiple computers in a transparent manner, these networks enable teams from different departments to access current and historical data with ease (Hin, 2020; Hong Hin, 2020). This level of accessibility empowers team members to detect potential errors or fraudulent activity without relying on centralized authorities (Figure 5.2). Another key benefit of distributed networks is immutability—once recorded on the blockchain chain, all transactional data becomes permanent and unalterable. Before being added to the chain itself, every block must be validated by all network participants while receiving its own unique timestamped record. As such, each piece of information can always be traced back through time, revealing where assets originated from as well as what processes they have undergone during their lifecycle, creating an unparalleled level of transparency throughout the entire supply chain process.

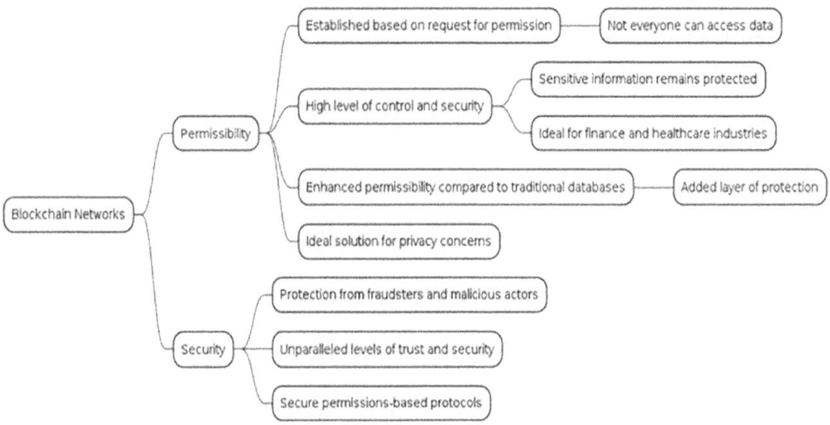

FIGURE 5.2 Defining features of block chain networks for financial systems.
Source: Author's Research.

One of the defining features of blockchain networks is their permissibility, which can be established based on a request for permission. Through this method, not everyone can access the data contained within these systems. This high level of control and security means that sensitive information remains protected from fraudsters or other malicious actors seeking to exploit it (Kim, 2019; Yamuna et al., 2023). By restricting access in this way, blockchain technology provides enhanced permissibility compared to traditional databases or digital storage methods. This added layer of protection makes it an ideal solution for industries such as finance or healthcare, where privacy concerns are paramount. Overall, by leveraging its unique ability to establish secure permissions-based protocols, blockchain offers unparalleled levels of trust and security in today's increasingly digitized world (Nakamoto, 2017; Sathya and Balakumar, 2022).

5.3 FRAUD DETECTION USING BLOCKCHAIN

Financial transactions are highly intricate and multifaceted processes that involve numerous parties, each with their own unique set of requirements. From collateral requirements to settlement times and currency differences, there are countless factors at play which contribute to the inherent complexity of these transactions (Hin, 2020; Hong Hin, 2020; Hu, 2022). In many cases, such processes may require multiple human interactions in order to be completed successfully. Fortunately, blockchain technology has revolutionized this landscape by enabling the detection of suspicious behaviors with ease. By leveraging blockchain's capacity for real-time data sharing and updates—approved only by those parties who have access—it becomes much easier for stakeholders involved in financial transactional activities to maintain transparency throughout every step of the process (Weingärtner et al., 2021).

Distribution fraud in financial markets has become a pervasive issue for companies across industries, causing significant financial and reputational damage (Hong Hin, 2020; Sathya and Balakumar, 2022; Weingärtner et al., 2021). With multiple distributors and companies involved in the process, it is difficult to maintain control over data access as the number of individuals with varying degrees of involvement increases exponentially. In such a complex ecosystem, ensuring transparency and accountability becomes crucial to prevent fraudulent activities (Figure 5.3).

FIGURE 5.3 Defining features of block chain infrastructure for financial systems.

Source: Author's Research.

Fortunately, blockchain technology presents an innovative solution that can address these challenges effectively. By providing businesses with transparent tracking capabilities that are secure from manipulation or tampering attempts, blockchain significantly enhances distribution chain integrity and helps mitigate fraud risks (Hin, 2020; Naik et al., 2020).

To achieve this level of security and reliability in product or service tracking using blockchain-based systems requires verification from all authorized participants before any updates are made to existing records or new entries added into them (Figure 5.4). This ensures an immutable audit trail that offers unparalleled visibility into every step along the distribution journey—eliminating discrepancies between distributors' claims about their products' origins while enhancing trust among stakeholders throughout the process (Barak et al., 2022; Hin, 2020). In conclusion, by leveraging blockchain's unique features, distribution fraud can be checked in time (Hong Hin, 2020; Hu, 2022).

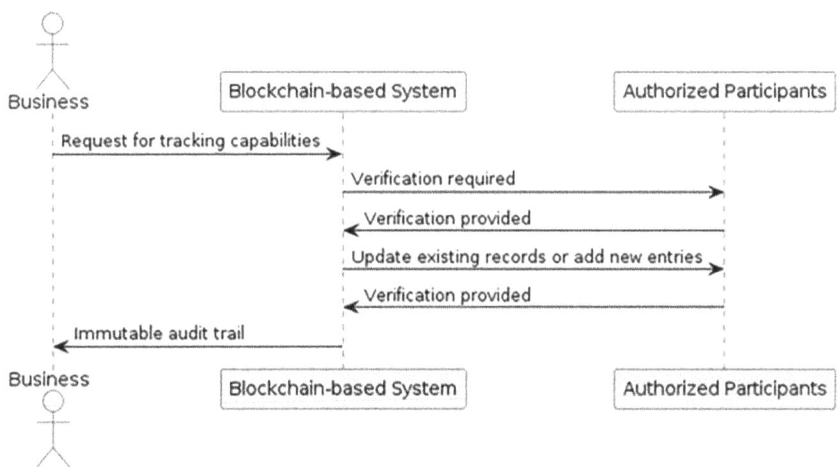

FIGURE 5.4 Complex financial ecosystem redefined by block chain.

5.3.1 BANKING FRAUD PROTECTION USING BLOCKCHAIN

In today's digital age, trust and confidence in fraud prevention and security technologies have become non-negotiable. It is expected that merchants employ top-notch fraud-proof solutions to protect their customers' financial data while also ensuring the safety of their own business operations (Mohamad and Al Harthy, 2022; Trad, 2019). On the other hand, customers are required to remain vigilant by always safeguarding their personal information (Liao and Wang, 2020). As transactions take place between issuers, acquirers, and various participants involved in payment processing activities, a myriad of layers exists within this secure environment—each playing an important role towards maintaining integrity throughout the process (Hong Hin, 2020). However, with these many layers comes an increased risk for potential errors or disruptions such as lost paperwork, deleted files, or even chargeback scams (Liu et al., 2022). What if there was a way to streamline these accountability measures? A decentralized approach where transactions can be stored securely without requiring human intervention could just be what we need—enter blockchain banking. By leveraging this innovative technology solution that uses cryptography (a form of encryption) for secure record keeping, it is kept secure (Liu et al., 2022; Sathya and Balakumar, 2022).

Blockchain is an innovative technology that utilizes a shared, secure ledger to keep track of and authenticate each component involved in a transaction (Sokolova et al., 2019; Weingärtner et al., 2021). This system involves the use of blocks which represent every step within the transactional process (Ark, 2018; Schifrin, 2019). These blocks are linked together securely as part of an unalterable chain, resulting in a detailed record for every transaction. One crucial aspect of blockchain's operation is its ability to timestamp each block accurately while also incorporating other identifying data elements such as originator information, time stamps, and more—this results in unparalleled transparency for everyone participating (Sokolova et al., 2019; Tayeb and Lago, 2018).

Furthermore, with blockchain transactions wrapped into one linear, stream-of-consciousness-style structure from start to finish, it eliminates multiple levels or layers that come with traditional review processes, leading towards greater efficiencies and cost savings (Barak et al., 2022; Yamuna et al., 2023). With this enhanced security protocol comes increased reliability across all applications—banking services like payment processing, contract management, real estate transfers—even personal medical records can be better protected by implementing Blockchain-based systems (Abdelhalim and Ibrahim, 2023; Charde and Hande, 2022; Yamuna et al., 2023).

5.3.2 ELIMINATING IDENTITY THEFT AND FRAUD

The fight against identity theft and fraud is an ongoing battle that continuously challenges all parties involved in the buying and selling process. Despite numerous advancements in security technology, vulnerabilities remain inherent within payment systems and data protection networks.

Unfortunately, hackers and fraudulent individuals have learned how to outsmart even the most sophisticated technological innovations (Figure 5.5).

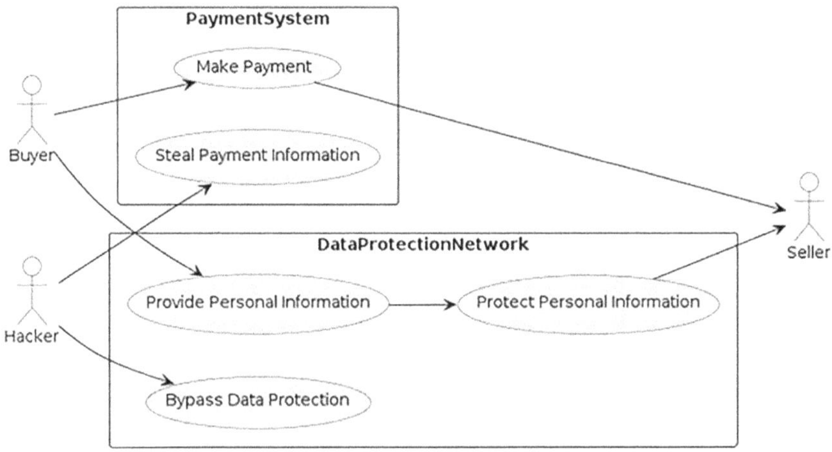

FIGURE 5.5 Eliminating process of identity theft and fraud detection.
Source: Author's Research.

Javelin Strategy and Research's 2017 Identity Fraud Study highlights a stark reality: In 2016 alone, over 15 million consumers fell prey to various forms of fraud or identity theft – marking a shocking increase of 16% from previous years. The report shows that card-not-present (CNP) fraud increased by nearly half, at around 40%, while account takeover fraud climbed up to almost one-third higher than before, at about 31%. Furthermore, new account scams were found rising steadily upwards at approximately 20%. The impact these increases had on businesses was catastrophic; losses due to fraudulent activities amounted to $16 billion in just one year (Mohamad and Al Harthy, 2022; Naik et al., 2020).

It is a growing concern amongst individuals across various sectors. Thankfully, due to its strong authentication systems, Blockchain offers unparalleled protection against fraudulent activity, as only authorized users can gain access while retaining personal safety due to protective security mechanisms embedded within its infrastructure (Yamuna et al., 2023). With certain aspects being handled exclusively through verification by trusted parties/specific authorized personnel, it provides protection for financial and identity fraud (Abdelhalim and Ibrahim, 2023).

Blockchain finance has the potential to revolutionize how data is created, approved, transferred, and stored by providing a level of depth that traditional systems cannot match (Inayatulloh et al., 2022; Li, 2022; Yang, 2022). The single ledger system employed by blockchain banks enables them to eliminate multiple layers of data transfer that occur within a transaction. This not only reduces room for error but also mitigates risks associated with time delays and vulnerability. By utilizing timestamps, blockchain ensures real-time monitoring, which fixes traceable timelines indicating accountability in transactions, including who was involved, when it took place, and where it happened. Every block or component of an account on the blockchain network is linked to an individual or organization's actual identity, making anonymity impossible (Naik et al., 2020).

Moreover, third-party approval is eliminated in this innovative technology, as each transaction requires 51% approval from those directly involved instead of relying on external parties for verification (Naik et al., 2020). Only individuals connected through the chain can access and approve these transactions, ensuring greater transparency throughout each stage of operation (Figure 5.6).

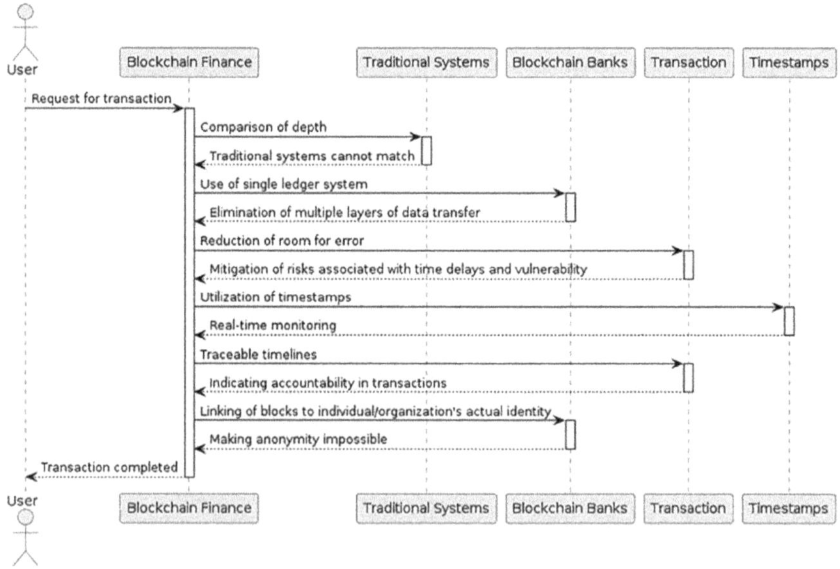

FIGURE 5.6 Process and determinants of secure block chain finance.

Source: Author's Research.

5.3.3 LEDGER-BASED SYSTEM OF INFORMATION FLOW

In today's digital age, paper receipts and transaction records are becoming a thing of the past as they are rapidly being replaced by secure blockchain technology. With this innovative solution, proof of purchase, approval records, receipt of items, and all payment data can be easily stored in an immutable ledger that is accessible to all stakeholders, including merchants, issuers, acquirers, and customers alike. This not only saves time but also reduces costs significantly while providing an added layer of security against fraudsters who try to insert fake or fraudulent information into the system. As everything is interlinked within the blockchain process itself, errors like double purchases or mistakes in approvals become nearly impossible to occur. The real advantage here lies in how such a fool-proof chain-of-data can change how information exchanges take place across different levels: alleviating pressures from various participants involved in any transaction, thereby creating more secure communication channels for everyone concerned. By taking away the burden on each participant involved during a sale/purchase scenario, risk is eliminated (Abdelhalim and Ibrahim, 2023; Ark, 2018).

The world of technology is constantly evolving, and while the emergence of blockchain banking has brought with it numerous advantages such as decentralized data storage and majority approval for every block, there remains a palpable element of risk (Kim, 2019; Tayeb and Lago, 2018). The Bitfinex hack in August 2016 serves as a stark reminder that despite its touted security measures, the system can still be infiltrated by savvy hackers who manage to gain control over vital keys and signatures which are crucial for ensuring utmost protection within the chain. Regrettably, this attack resulted in an unprecedented loss of $60 million worth of Bitcoin – funds which remain unrecovered to date – hence underscoring how critical it is for stakeholders in this emerging field to always prioritize comprehensive safety protocols.

With its unique ability to provide a decentralized and secure platform for transactions, blockchain technology has the potential to revolutionize the way we approach security and fraud prevention (Rao, 2020; Wang and Wang, 2022). However, much like the Internet during its early stages in the 1990s, this innovation is still very much a work in progress as it undergoes continuous development and improvement (Rao, 2020; Yang, 2022). While there is no denying that blockchain can offer numerous benefits for businesses looking to enhance their security measures, it's important not to get too caught up in hype surrounding this technology. As history has shown us with other innovative payment methods such as m-commerce, Tap and Go systems, or even smartwatches, change takes time.

5.3.4 INFORMATION FLOW USING BLOCKCHAIN.

At the forefront of blockchain implementation is the first and most basic level, where a company can securely store information on the distributed ledger without any involvement from third parties (Kim, 2019; Liao and Wang, 2020; Sathya and Balakumar, 2022; Tayeb and Lago, 2018). This unidirectional flow of data provides a reliable and transparent method for companies to record transactions and other vital information. However, taking it up a notch in level two involves incorporating an external party like a validator or counterparty who must interact with uploaded data by either confirming or validating inputs (Hu, 2022). The bidirectional flow that ensues allows for more efficient verification processes. For those seeking even greater depth in their blockchain implementations, smart contracts take center stage at level three (CEBECİ and ARAT, 2022; Nakamoto, 2017).

These software protocols are capable of automatic execution within the blockchain ecosystem without requiring intervention from outside entities. Herein lies not only bi-directional flows but also trigger actions independent of any interaction between involved parties – raising possibilities for far-reaching applications across diverse industries through automation-enabled transparency secured against tampering by cryptographical methods utilized throughout levels one to three (Figure 5.7).

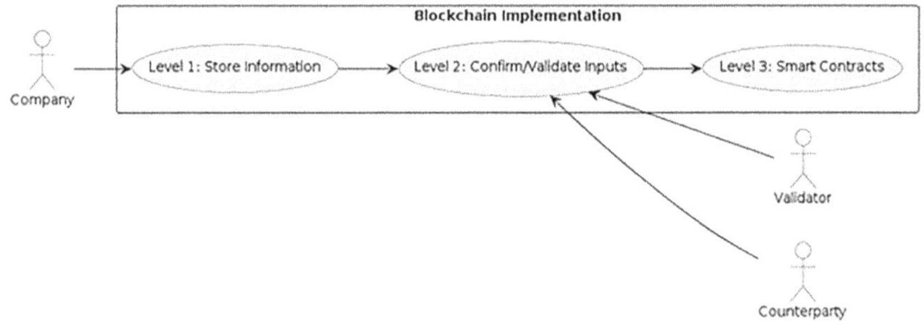

FIGURE 5.7 Information flow monitoring using block chain.

Source: Author's Research.

There are several case studies (Inayatulloh et al., 2022; Jameaba and Ssenyonga Jameaba, 2022; Wang and Wang, 2022) that showcase the incredible potential of blockchain technology in facilitating secure and efficient information flow. One particularly compelling application is its ability to detect instances of fraud with remarkable speed and accuracy, providing unprecedented levels of protection for individuals and organizations alike (Yang, 2022). Through robust encryption protocols, decentralized distribution mechanisms, and real-time monitoring capabilities, blockchain offers an unparalleled level of transparency and accountability that makes it ideally suited to combatting fraudulent activities in a wide variety of contexts (Figure 5.8). Whether applied within financial services platforms, this innovative technology promises transformative benefits for all those who seek reliable data security solutions (Vivekanadam, 2020).

FIGURE 5.8 Influence of secure block chain finance on financial systems.

Source: Author's Research.

Furthermore, this technology enables real-time verification of information sharing between stakeholders, which further enhances accountability while reducing human error (Liu et al., 2022; Schifrin, 2019). With such benefits offered by blockchain technologies, it is no surprise that it is increasingly being integrated into many business models with tremendous success. Banks and financial institutions utilizing such solutions are better equipped at dealing with cybercrime threats related to hacking or unauthorized access attempts on their databases, providing clients peace of mind knowing their assets remain safeguarded from malicious attacks (Li, 2022; Yang, 2022).

5.4 ANALYZING FINANCIAL HISTORY AND BACKGROUND

In today's digital age, the banking industry has experienced an increased number of loan fraud cases due to falsified customer information (Charde and Hande, 2022). The traditional method of assessing a customer's creditworthiness involves analyzing personal financial data such as repayment history and income level. However, with online application systems becoming more prevalent, customers may be tempted to provide misleading information to increase their chances of securing a loan. This deceitful practice can result in significant losses for financial institutions, which ultimately impacts the wider economy (Charde and Hande, 2022; Hong Hin, 2020). For this reason, blockchain technology has emerged as a potential solution that could revolutionize how banks assess risk during the lending process (Maeng, 2017; Zelic and Baros, 2018). By leveraging decentralized ledgers that securely record historical transaction records across multiple nodes on the network, applicants are unable to manipulate or tamper with any aspect of their personal information submitted through online applications without being detected by blockchain monitoring protocols (Li, 2022; Xia et al., 2022). This not only reduces incidences of fraudulent activity but also enables lenders to make informed decisions based on accurate data, thereby mitigating risk and improving operational efficiency within the sector (Inayatulloh et al., 2022; Wang and Wang, 2022).

Also, blockchain technology provides a decentralized system that allows for secure transactions without requiring intermediaries such as banks or governments. By utilising blockchain, one can track financial activity with greater depth than ever before (Hong Hin, 2020; Yamuna et al., 2023). This includes analyzing past transactions and tracking current ones in real-time. In addition, data transparency on the blockchain platform enables individuals and businesses alike to verify any transactional details securely (Schifrin, 2019). The significance of understanding an entity's financial history cannot be overstated; it may provide valuable insights into its operations, performance, risk management practices, and so forth, which can aid investors while making informed decisions regarding investments. Thus, using a robust blockchain infrastructure to assess an individual or business' economic activities can help increase accuracy while adding layers of security required in these modern times where fraudsters are always looking for loopholes within traditional financial systems (Maeng, 2017).

In various case studies, blockchain has demonstrated its potential to revolutionize traditional methods of financial analysis through greater transparency, immutability, and security (Barak et al., 2022; Sokolova et al., 2019; Tayeb and Lago, 2018; Trad, 2019). With blockchain-enabled systems, analysts can gain access to a comprehensive database that tracks every transaction made within the system's network (Nakamoto, 2017). This allows for an in-depth examination of past transactions while maintaining an unalterable record trail that provides insight into areas such as payment histories and investment portfolios (CEBECİ and ARAT, 2022). Furthermore, because each block on the chain cannot be tampered with or deleted without affecting all subsequent blocks' integrity, this creates an additional layer of security against fraudulent activity. Overall, these developments suggest that blockchain can offer financial regulators new ways to investigate financial history – from banking crises to monetary policy decisions – enabling deeper insights into actions (Charde and Hande, 2022; Kim, 2019; Zelic and Baros, 2018).

However, before fully embracing blockchain solutions en masse across various financial systems, thorough testing needs to be done because, while promising right now, it remains an emerging solution among many others trying to solve some of our most pressing issues related to data protection, security breaches, and fraudulent activities online.

5.4.1 BLOCK CHAIN FOR FINANCIAL RECORD KEEPING

In today's digital age, where data breaches and fraudulent activities are increasingly common, ensuring secure recordkeeping is crucial (Hong Hin, 2020). Blockchain technology provides a unique solution to this problem by creating an incorruptible ledger of transactions that can be accessed by all parties involved in the network. The foundation of the blockchain lies in its ability to interlink each transaction with previous ones, forming an unbreakable chain of records (Hong Hin, 2020; Yamuna et al., 2023). This process requires group consensus amongst all participants on what should be deemed as the most recent version of the blockchain; hence, adding new transactions becomes complex and costly. To ensure validity and acceptance for each new entry into the system, every node needs to perform computations using hash functions, making it extremely challenging for any party or individual to manipulate previously recorded information. Once verified and included within the ledger, these entries become permanent fixtures that cannot be easily altered or deleted, resulting in an irreversible chain-of-events timeline (Charde and Hande, 2022; Hong Hin, 2020).

The potential of this innovative system has been recognized by asset management organizations who are utilizing it to maintain meticulous records that ensure security for investors' valuable assets. By leveraging the transparency and tamper-proof characteristics of Block Chain, these companies can guarantee an accurate trail of all financial transactions without the need for intermediaries or third-party verification processes. This not only enhances trust between stakeholders but also creates more efficient and cost-effective methods of managing finances while providing complete peace of mind regarding data integrity and protection against fraudulence.

Several case studies provide evidence of the benefits of using blockchain for financial record-keeping purposes (Ark, 2018; Rao, 2020). For instance, some banks are implementing private blockchains for trade finance activities such as letter of credit processing or supply chain financing operations that require secure data sharing across multiple parties while maintaining confidentiality (Figure 5.9). Another example comes from cryptocurrency exchanges which rely on public blockchains to facilitate trading by recording all transactions onto immutable ledgers instantly accessible by anyone who has access to them. This not only improves transparency but also provides instant verification, thereby reducing settlement times and risks associated with frauds like double-spending attacks. Moreover, since it's almost impossible to manipulate records stored on Blockchain networks once they're entered the system due to its nature.

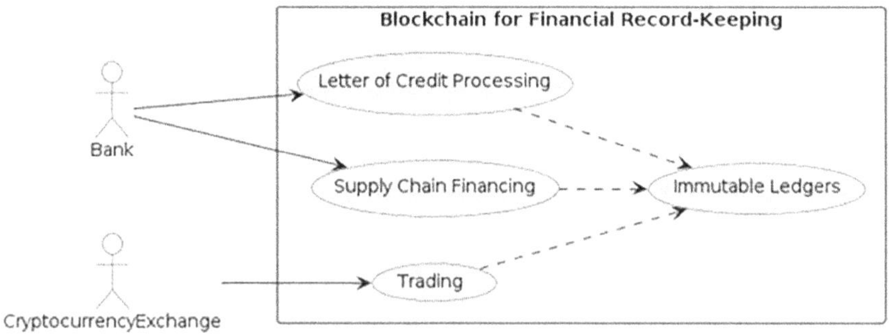

FIGURE 5.9 Influence of secure block chain on financial record keeping.

Source: Author's Research.

5.4.2 FINANCIAL DATA PRIVACY AND ACCESSIBILITY CONCERNS

In the realm of data privacy, there has been a growing concern among portfolio users who are hesitant to provide sensitive information. As a solution to this pressing issue, numerous case studies have emerged that explore how blockchain technology can be leveraged for secure and reliable reputation systems. To achieve maximum protection of users' personal data, many researchers propose utilizing digital signatures in conjunction with blockchain networks. By employing public and private keys within these systems, both parties involved can enhance accessibility while simultaneously upholding stringent security measures (Figure 5.10).

Overall, it is evident that blockchains offer tremendous potential when it comes to safeguarding the integrity and confidentiality of financial information in today's digital age. Further advancements are being made each day by leading experts across various industries; more innovative applications are being applied using blockchain technology for heightened levels of security moving forward.

There are numerous banks and financial institutions that have already started utilizing the benefits of blockchain technology. This revolutionary technology has proved to be a game-changer in the world of finance, enhancing security measures, promoting transparency, reducing costs, and improving efficiency across various sectors. Some notable examples include leading banks such as JP Morgan Chase, Citibank, and Wells Fargo, who have embraced blockchain for their payment systems. Additionally, Santander is using this cutting-edge technology for cross-border

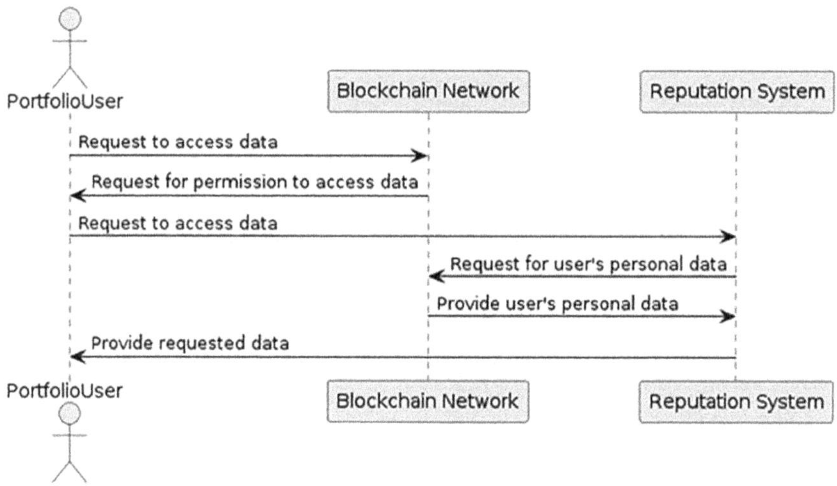

FIGURE 5.10 Secure block chain affecting financial data privacy and accessibility concerns.

Source: Author's Research.

payments, while Goldman Sachs is incorporating it into its clearing services. Furthermore, some central banks around the globe, including the Bank of England and the Monetary Authority of Singapore, have launched successful pilot projects on their respective digital currencies built on top of blockchain platforms. Even traditional stock exchanges like Nasdaq are exploring potential applications with regards to trade settlements through smart contracts implemented via blockchain. These instances only depict a few among many developments in which enterprises from different sectors are successfully leveraging blockchain's capabilities to enhance their offerings towards customers whilst unlocking new business opportunities at the same time.

5.4.3 CREATING REPUTATION SYSTEMS AND FEEDBACK MECHANISMS

As the world of finance continues to evolve, an increasing number of banks and financial institutions are turning to blockchain technology as a means of creating reputation systems. These innovative systems leverage the power and security of blockchain to provide more robust, trustworthy methods for evaluating individuals and entities within the industry. One such example is JP Morgan Chase, which has created its own proprietary blockchain platform known as Quorum. This system enables the development of custom reputation management tools that can be tailored specifically to individual clients' needs. Additionally, Quorum boasts advanced privacy features that allow sensitive information about users to be securely managed without sacrificing data integrity or accessibility. Another institution making use of blockchain-based reputation systems is Goldman Sachs. The investment bank has partnered with fintech firm Circle on developing CENTRE Consortium – a network designed explicitly around facilitating efficient cross-border transactions through utilizing stablecoins (digital currencies pegged against assets). By using open standards built atop blockchains like Ethereum; businesses from any country can seamlessly transact globally.

Numerous banks and financial institutions have recently embraced the revolutionary blockchain technology to create feedback mechanisms for their investors and account holders. The implementation of this cutting-edge technological advancement has enabled them to enhance the depth, accuracy, and efficiency of these feedback systems substantially. Through utilizing blockchains in designing such advanced mechanisms, these establishments can deliver more secure, transparent, tamper-proof data records that provide a comprehensive insight into clients' behaviors. This allows stakeholders to make informed decisions based on an exhaustive analysis of each transaction's intricate details while keeping all parties involved in check with reliable verifications processed within seconds rather than days or weeks as previously required by traditional methods. It enables instant verification of data updates across multiple parties, eliminating any possibility for discrepancies or errors. Some leading names in this area include JP Morgan Chase Bank, which developed its own digital currency called JPM Coin as part of an initiative to enhance payment processing efficiency using blockchain; CitiBank's creation of Citicoin, as well as Barclays' pilot program aimed at simplifying identity checks and robust feedback evaluations via distributed ledgers, amongst others (Figure 5.11). Overall, adopting Blockchain within financial operations is not just about improving customer experience but also serves as a significant step forward for more efficient management practices.

FIGURE 5.11 Creating reputation systems and feedback mechanisms using block chain.
Source: Author's Research.

5.4.4 ELIMINATION OF RATING FRAUD IN FINANCIAL SYSTEMS

Rating fraud can manifest in various types of reputation systems, each presenting unique challenges. While blockchain technology effectively verifies factual information and content-driven ratings in non-computational systems, user-driven reputation models pose a more complex problem. Ratings provided by individuals are inherently subjective and often lack concrete evidence or proof to support them. Even if an attacker manipulates ratings unfairly to discredit a particular item, they may argue that their opinion was merely influenced by personal preference rather than objective criteria, making fraudulent behavior harder to detect even with extensive historical data records available. To address this issue, Dennis and Owenson (2016) have suggested removing human opinions from these reputation systems entirely. This involves creating fully automated processes for determining credibility without any input from users themselves, ultimately streamlining the system's accuracy while also mitigating opportunities for rating fraud to occur. This approach emphasizes an objective representation of reputation through binary values, reflecting whether users received the file or not. By limiting subjective input from humans, this system prioritizes fact-based information over personal biases or preferences. Furthermore, leveraging blockchain technology can enhance fraud detection capabilities within these objectively driven reputation systems by providing increased security and transparency in data handling processes.

5.4.5 WEAPON AGAINST WHITEWASHING ATTACKS USING BLOCKCHAIN

The utilization of blockchain technology is proving to be a promising weapon against the damaging and pervasive threat of whitewashing attacks. With reputation systems built on blockchain, user accounts can be established with real identities that remain protected from disclosure. This innovative approach ensures that individuals attempting to falsify subjective information are unable to simply abandon their account to create a new one and thus evade accountability for their past actions. By rendering such efforts futile, this cutting-edge technology delivers greater security and transparency than ever before seen in combating fraudulent activities online.

Blockchain-based reputation systems offer an innovative solution by allowing user accounts to be created with real identities, while maintaining anonymity and privacy. This means that even though users' true identity remains concealed from other participants, it can still be verified through smart contracts on the blockchain. As a result, once an unscrupulous rater injects fraudulent subjective information into the system to manipulate ratings for personal gain, they cannot simply leave the platform and create new accounts under different aliases or pseudonyms to cover their tracks—something which has been traditionally very difficult, if not impossible, in many online rating systems. Instead, any malicious activity will remain linked forever to their account, making attempts at whitewashing or concealing previous negative behavior futile since all historical data recorded on public ledgers are immutable. The impact of this innovation promises far-reaching consequences for enhancing trustworthiness within various financial systems where accurate reputational data is much valued.

5.4.5 BLOCKCHAIN-BASED FINANCIAL PAYMENT AND TRANSACTION SYSTEMS

One example of blockchain-based payments is BitPesa, a company that operates in several African countries like Kenya, Nigeria, and Uganda. This innovative payment platform has facilitated millions of dollars' worth of transactions and shows no signs of slowing down with a reported 20% month-over-month growth rate. Interestingly, BitPesa has made significant strides towards reducing the high costs associated with remittances sent throughout sub-Saharan Africa—reportedly the most expensive region in the world for sending money. The adoption of crypto payments platforms such as BitPesa have led to an incredible reduction (over 90%) in transfer fees within this region alone. Another prominent player within cryptocurrency payment services is BitPay, which aids merchants by enabling them to accept and store bitcoin payments securely while providing integrations with e-commerce platforms including Shopify and WooCommerce, among others. In recent years, DoorDash alongside Uber Eats have also been added to their list, resulting in yet another convenient option for making purchases using cryptocurrencies.

The Ethereum-based payments platform AirFox has recently taken its customer convenience to the next level by partnering with MasterCard. Thanks to this strategic collaboration, customers can now use the popular banQi app at a global network of points-of-sale and every Via Varejo location. In another part of the world, South Korea-based crypto payments startup HUPAYX has been making waves since 2019 when it partnered with several local businesses to establish an extensive payment network. This allowed consumers in South Korea access to over 400,000 stores, including duty-free outlets and shopping complexes, where they could pay conveniently using their mobile devices through HUPAYX's point-of-sale infrastructure (Jameaba and Ssenyonga Jameaba, 2022; Xia et al., 2022). Meanwhile, companies such as Bitwage and PaymentX have revolutionized traditional payroll systems by enabling employers worldwide to offer employees cryptocurrency payments instead of more conventional methods. With innovative services like these paving the way for digital currencies' mainstream adoption across different industries globally, experts predict that cryptocurrency is set for even greater success soon (Yeasmin and Baig, 2019; Zhu et al., 2020).

Blockchain also facilitates micropayments with its decentralized nature and inherent security features. This technological innovation is empowering more secure and efficient payment methods to emerge, making it easier for individuals to access premium online content even for small purchases that were previously not viable options due to high transaction costs associated with traditional banking institutions' fee structures. For instance, SatoshiPay is an exceptional example of how blockchain-powered micropayment systems work seamlessly. It offers an online cryptocurrency wallet where users can easily pay-per-view using minute amounts from their wallets loaded with various supported payment tokens such as bitcoin or US dollars. This way, accessing paid content becomes much easier and cost-effective than ever before. With these advancements in blockchain technology's application towards facilitating microtransactions on platforms like SatoshiPay, there is a paradigm shift in how people transact value digitally today.

Ripple, an enterprise blockchain services provider, is making waves in the finance industry as it works towards building out innovative solutions for banks to use in clearance and settlement. Despite being best known for its associated cryptocurrency XRP, Ripple's focus on leveraging blockchain technology has allowed them to remove intermediaries from asset rights transfers, which results in significantly lower exchange fees while also providing access to a wider range of global markets. Additionally, using this new method reduces the instability present within traditional securities markets. As more companies raise funds through initial coin offerings (ICOs) powered by public blockchains like Ethereum and Bitcoin, there is an increasing potential for widespread adoption of blockchain across various sectors. By moving securities onto these secure and immutable ledgers that are available via public blockchains, it is projected that $17 billion–$24 billion per year could be saved globally just in trade processing costs alone.

There is a growing interest among blockchain technology companies in facilitating the migration of trillions of dollars' worth of financial securities onto the blockchain. A prime example is Polymath, which launched its own bespoke blockchain network—Polymesh—in 2021 with regulatory approval from 14 major financial entities as node operators to verify and validate new blocks added to their chain. Furthermore, they have also brought on board investment firm Greentrail and digital asset exchange Huobi as additional node operators for enhanced credibility.

The potential benefits are not lost on traditional stock exchanges either; The Australian Stock Exchange plans to replace its current system for book-keeping, clearance, and settlements with Digital Asset Holdings' cutting-edge proprietary blockchain solution by April 2023. Across the pond, the London Stock Exchange has recently hired a head of group digital assets, signaling that it too may be leaning towards adopting similar systems soon enough. The announcement followed closely behind the UK government's revelation about introducing legislation intended at promoting financial innovation around blockchain.

5.5 CONCLUSION

It is crucial to acknowledge that while blockchain systems have revolutionized recordkeeping with their ability to maintain permanent and unalterable records, they are not infallible when it comes to information security in financial systems (Fiergbor, 2019). Despite the immutable nature of these systems, certain vulnerabilities exist which can be exploited by tech-savvy hackers for malicious purposes. For instance, sophisticated cyber attackers can infiltrate blockchain networks and inject corrupted nodes or spread deadly viruses across all computers in the network. As a result of this contagion effect, even more devices may become infected at an alarming rate, leading to disastrous consequences if left unchecked (Hong Hin, 2020). With the implementation of blockchain systems, there is an inherent protection of users' privacy, which grants greater confidence in the security and integrity of submitted data. To ensure that ratings are authentic, it is necessary to restrict submissions solely from accounts created using real identities. By doing so, this eliminates fraudulent practices commonly observed within traditional reputation systems where a single individual could control multiple account IDs and inject falsified ratings (Ren et al., 2020).

Although blockchain technology has been hailed as a promising solution for mitigating fraud in financial systems, some experts have raised concerns regarding its effectiveness. While it is true that blockchain offers many benefits such as decentralization and transparency, there are still challenges that need to be addressed before it can fully eliminate fraudulent activities. One example of this is the issue of scalability (Hu, 2022; Nakamoto, 2017). As more transactions are added to the network, the time needed for validation increases significantly, leading to potential bottlenecks or delays which could compromise security measures already put in place. Furthermore, while data stored on a blockchain may be immutable once recorded, they do not necessarily guarantee authenticity, thereby exposing loopholes where malicious actors might exploit vulnerabilities within blockchain networks by injecting fake transaction records into the system. The benefits offered by blockchain-based reputation systems transcend beyond prevention against fraudulence as they promote transparency and accountability while simultaneously decreasing trust barriers between parties involved. Despite limitations, however, with continued research and development coupled with robust regulatory frameworks aimed at ensuring strict compliance requirements across all sectors using blockchain technology, eliminating fraud from our global financial infrastructure remains an attainable goal worth pursuing relentlessly until success.

Further research exploration into blockchain technology is required for future advancements such as detecting both objective as well as subjective financial information frauds, thus enhancing our overall understanding of this innovative field.

KEYWORDS

- **blockchain**
- **data privacy**
- **financial ecosystem**
- **financial systems**
- **fraud detection**
- **information flow**
- **ledger**
- **theft and fraud**

REFERENCES

Abdelhalim, A. M. M., & Ibrahim, N. M. E. S. (2023). The impact of using smart algorithms and blockchain technology on the profits' quality in Saudi financial market. *Lecture Notes in Networks and Systems*, *557*. https://doi.org/10.1007/978-3-031-17746-0_34 (accessed on 26th July 2025).

Ark, T. (2018). 20 ways blockchain will transform (okay, may improve) education. *Forbes*. https://www.forbes.com/sites/tomvanderark/2018/05/14/20-ways-blockchain-will-transform-okay-may-improve-education/ (accessed on 26th July 2025).

Barak, J., Chaudhary, H., Mangal, R., Goel, A., & Sharma, D. K. (2022). Blockchain in smart healthcare management. In *Internet of Healthcare Things: Machine Learning for Security and Privacy*. https://doi.org/10.1002/9781119792468.ch8 (accessed on 26th July 2025).

CEBECİ, U., & ARAT, E. (2022). Blockchain'e dayalı tarım ve gıda tedarik zinciri kaynağının kurulması: Literatür incelemesi. *European Journal of Science and Technology*. https://doi.org/10.31590/ejosat.1131779 (accessed on 26th July 2025).

Charde, E., & Hande, K. (2022). Assessment of usefulness of blockchain technology for agriculture application. *International Journal for Research in Applied Science and Engineering Technology, 10*(5). https://doi.org/10.22214/ijraset.2022.43031 (accessed on 26th July 2025).

Fiergbor, D. D. (2019). Blockchain technology in fund management. *Communications in Computer and Information Science, 899*. https://doi.org/10.1007/978-981-13-2035-4_27 (accessed on 26th July 2025).

Hin, L. H. (2020). Critical review: Future of blockchain economy in financial services. *Global Journal of Management and Business Research*. https://doi.org/10.34257/gjmbrbvol19is7pg9 (accessed on 26th July 2025).

Hong Hin, L. (2020). Critical review: Current research issues on cryptocurrency and its application in financial sectors. *International Journal of Business and Management, 15*(3). https://doi.org/10.5539/ijbm.v15n3p139 (accessed on 26th July 2025).

Hu, Z. (2022). Application analysis of block technology based on blockchain computer model in the false identification of the catalog list of the financial department of multinational companies. *Advances in Multimedia, 2022*. https://doi.org/10.1155/2022/8193204 (accessed on 26th July 2025).

Inayatulloh, Onsardi, Suwarni, E., Mangruwa, R. D., Djajasinga, N. D., Darmawati, & Nuryadin, B. (2022). Increasing efficiency and transparency of soft loans for SME businesses with blockchain technology. *International Journal of Applied Engineering and Technology (London), 4*(2).

Jameaba, M. S., & Ssenyonga Jameaba, M. (2022). Digitalization, emerging technologies, and financial stability: Challenges and opportunities for the banking industry. *Qeios, Ic*.

Kim, D. (2019). Digital economy and FinTech in the view of consumer protection. *The Journal of Comparative Private Law, 26*(3).

Li, Y. (2022). Analysis of financial sharing model based on blockchain technology in internet + 5G environment. *Lecture Notes on Data Engineering and Communications Technologies, 123*. https://doi.org/10.1007/978-3-030-96908-0_100 (accessed on 26th July 2025).

Liao, Q., & Wang, Y. (2020). Prospect and challenges of cross-border payment posed by digital currency – From the perspective of blockchain coalition. *E3S Web of Conferences, 218*. https://doi.org/10.1051/e3sconf/202021804001 (accessed on 26th July 2025).

Liu, J., Yan, L., & Wang, D. (2022). A hybrid blockchain model for trusted data of supply chain finance. *Wireless Personal Communications, 127*(2). https://doi.org/10.1007/s11277-021-08451-x (accessed on 26th July 2025).

Maeng, S. S. (2017). Rational regulation plans for blockchain-based virtual currency – Focused on Bitcoin. *Commercial Law Review*, *35*(4). https://doi.org/10.21188/clr.35.4.5 (accessed on 26th July 2025).

Mohamad, M., & Al Harthy, K. (2022). Proposed architecture of integrate blockchain with VAT system in Oman. *International Conference on Electrical, Computer, and Energy Technologies, ICECET 2022*. https://doi.org/10.1109/ICECET55527.2022.9872925 (accessed on 26th July 2025).

Naik, V., Pejawar, R., Singh, R., Aher, A., & Kanchan, S. (2020). Expeditious banking using blockchain technology. *International Conference on Computational Intelligence for Smart Power System and Sustainable Energy, CISPSSE 2020*. https://doi.org/10.1109/CISPSSE49931.2020.9212253 (accessed on 26th July 2025).

Nakamoto, S. (2017). Bitcoin for merchants: Legal considerations for form of payment. *Asia Pacific Journal of Innovation and Entrepreneurship*, *19*(1).

Rao, V. S. (2020). Blockchain accounting in a triple entry system–Its implications on the firm. *Our Heritage*, *30*.

Ren, Y. L., Xu, D. T., Zhang, X. P., & Gu, D. W. (2020). Scheme of revisable blockchain. *Ruan Jian Xue Bao/Journal of Software*, *31*(12). https://doi.org/10.13328/j.cnki.jos.005894 (accessed on 26th July 2025).

Sathya, M., & Balakumar, B. (2022). Insurance fraud detection using novel machine learning technique. *International Journal of Intelligent Systems and Applications in Engineering*, *10*(3).

Schifrin, A. (2019). AI startups and the fight against online disinformation. *German Marshall Fund of the United States*.

Sokolova, I., Orlovtseva, O., Leskina, O., Gubanova, E., & Kanikhin, T. (2019). Application of artificial intelligence capabilities for practical needs of participants in economic processes. *Proceedings of the 33rd International Business Information Management Association Conference, IBIMA 2019: Education Excellence and Innovation Management through Vision 2020*.

Tayeb, & Lago. (2018). Blockchain technology: Between high hopes and challenging implications. *The MENA Business Law Review*, *1*(2).

Trad, A. (2019). *The Business Transformation and Enterprise Architecture Framework*. https://doi.org/10.4018/978-1-7998-1279-1.ch019 (accessed on 26th July 2025).

Vivekanadam, B. (2020). Analysis of recent trend and applications in blockchain technology. *Journal of ISMAC*, *2*(4). https://doi.org/10.36548/jismac.2020.4.003 (accessed on 26th July 2025).

Wang, L., & Wang, Y. (2022). Supply chain financial service management system based on blockchain IoT data sharing and edge computing. *Alexandria Engineering Journal*, *61*(1). https://doi.org/10.1016/j.aej.2021.04.079 (accessed on 26th July 2025).

Weingärtner, T., Batista, D., Köchli, S., & Voutat, G. (2021). Prototyping a smart contract based public procurement to fight corruption. *Computers*, *10*(7). https://doi.org/10.3390/computers10070085 (accessed on 26th July 2025).

Xia, X., Hou, T., Liu, X., Zong, C., & Mu, S. (2022). Protecting check-in data privacy in blockchain transactions with preserving high trajectory pattern utility. *Wireless Communications and Mobile Computing*, *2022*. https://doi.org/10.1155/2022/9358531 (accessed on 26th July 2025).

Yamuna, S., Sakthi, U. S., Priya, G. S., & Sendhilkumar, K. S. (2023). Blockchain-based roaming fraud prevention using LSTM model in 4G LTE network. *Proceedings of the 13th International Conference on Cloud Computing, Data Science and Engineering, Confluence 2023.* https://doi.org/10.1109/Confluence56041.2023.10048873 (accessed on 26th July 2025).

Yang, R. (2022). Development and supervision of financial technology based on blockchain. *Computational Intelligence and Neuroscience, 2022.* https://doi.org/10.1155/2022/2615153 (accessed on 26th July 2025).

Yeasmin, S., & Baig, A. (2019). Unblocking the potential of blockchain. *2019 International Conference on Electrical and Computing Technologies and Applications, ICECTA 2019.* https://doi.org/10.1109/ICECTA48151.2019.8959713 (accessed on 26th July 2025).

Zelic, D., & Baros, N. (2018). Cryptocurrency: General challenges of legal regulation and the Swiss model of regulation. In *Economic and Social Development (ESD 2018): 33rd International Scientific Conference on Economic and Social Development–Managerial Issues in Modern Business.*

Zhao, J. (2021). Efficiency of corporate debt financing based on machine learning and convolutional neural network. *Microprocessors and Microsystems, 83.* https://doi.org/10.1016/j.micpro.2021.103998 (accessed on 26th July 2025).

Zhu, Q., Liu, B., Han, F., & Lee, M. (2020). The optimization effect of fuzzy fractional-order ordinary differential equation in blockchain financial cross-border e-commerce payment mode. *Alexandria Engineering Journal, 59*(4). https://doi.org/10.1016/j.aej.2020.06.031 (accessed on 26th July 2025).

CHAPTER 6

Evolution of FinTech Under Blockchain Technology

NISHA TOKAS,[1,*] NEHA PURI,[2] and AYUSHI TOMAR[1]

[1]*Maharaja Surajmal Institute, New Delhi, Delhi, India*

[2]*Amity College of Commerce & Finance (ACCF), Amity University, Noida, Uttar Pradesh, India*

*Corresponding author

ABSTRACT

Blockchain technology, renowned as the foundational technology behind cryptocurrencies, has ignited substantial interest across numerous industries due to its potential to revolutionize conventional methods and systems. In the financial realm, blockchain's impact is massive, with one sector poised for transformation being the derivatives market. The derivatives market holds a pivotal role in global finance, enabling participants to manage risks, speculate on asset prices, and hedge against market uncertainties. However, this market faces challenges, including complicated and fragmented structures, opacity, counterparty risk, and high transaction charges. Blockchain technology offers promising solutions to address these issues, with its integration into the derivatives market already underway. The primary impact of blockchain on derivatives lies in settlement and clearing processes. Conventional systems involve multiple intermediaries, leading to delays, expanded operational costs, and exposure to counterparty risks. Blockchain's distributed ledger technology allows for real-time and immutable transaction recording, notably reducing settlement times and enhancing transparency. As the technology matures and regulatory frameworks evolve, the derivatives market is poised for further transformation.

FinTech and Blockchain Technology in Financial Markets. Neha Puri, Anjali Munde, Emilia Alaverdov (Eds.)
© 2027 Apple Academic Press, Inc. Co-published with CRC Press (Taylor & Francis)

DOI: 10.1201/9781779641298-6

The potential for smart contracts, asset tokenization, and DeFi derivatives markets opens up exciting possibilities for more accessibility, efficiency, and innovation in this important area of global finance.

6.1 INTRODUCTION

The derivatives marketplace, a critical factor of worldwide finance, performs a fundamental role in managing risk, providing liquidity, and facilitating price discovery. Over time, the financial market has become more complex, presenting a wide variety of financial instruments, including futures, options, swaps, and forwards. However, the derivatives marketplace has not been resistant to challenges, including inefficiencies, opacity, counterparty risks, and high transaction fees. These days, the emergence of blockchain technology (Ali et al., 2015) has captured the attention of the economic enterprise, promising transformative solutions to these longstanding issues. This introduction explores the outcomes of blockchain technology on the derivatives marketplace, studying its potential to revolutionize conventional practices and reshape the landscape of financial derivatives.

Blockchain technology, initially popularized as the backbone of cryptocurrencies like Bitcoin, has evolved into a versatile and disruptive force across various sectors (Bitcoinwiki, 2015). Its decentralized, immutable, and transparent nature has drawn interest from financial institutions and regulators seeking to enhance security, transparency, and performance in financial transactions (Andrychowicz et al., 2015). With the aid of creating a decentralized and tamper-proof distributed ledger, blockchain eliminates the need for intermediaries, streamlines methods, and introduces new opportunities for financial innovation (Antonopoulos, 2014). In the context of the derivatives marketplace, blockchain generation gives a promising pathway to triumph over current challenges. One of the number one influences of blockchain on derivatives lies in agreement and clearing approaches. Traditional settlement systems often contain lengthy methods and multiple intermediaries, ensuing in delays, extended operational fees, and heightened counterparty risks. In evaluation, blockchain's (Ateniese et al., 2014) distributed ledger technology permits for real-time and immutable recording of transactions, drastically reducing settlement times and growing transparency. This transformation paves the manner for more efficient, cost-powerful, and secure derivatives buying and selling.

Moreover, blockchain technology introduces the idea of smart contracts, which own the capability to revolutionize the negotiation, execution, and agreement of by-product contracts (Swan, 2015). Smart contracts are self-executing agreements that automatically cause predefined conditions whilst certain events arise. By using leveraging smart contracts, participants can streamline the complete lifecycle of by-product contracts, from inception to settlement, with out the need for intermediaries. The automation (Decker & Wattenhofer, 2014) furnished by way of clever contracts minimizes the potential for human errors and disputes, ensuing in a more dependable and streamlined derivatives (Nakamoto, 2008) buying and selling manner.

Past operational enhancements, blockchain technology (Decker et al., 2015) also addresses the issue of transparency and regulatory compliance within the derivatives market. The immutable nature of blockchain facts guarantees that each one transaction details are completely saved and auditable. Regulators can get admission to a complete view of marketplace activities, thereby strengthening regulatory oversight. This more suitable transparency no longer handiest mitigates the risk of market manipulation and fraud but also instills greater self-belief amongst market individuals, fostering trust and credibility in the derivatives marketplace.

In addition to its transformative impact on existing tactics, blockchain generation has the capacity to enable fractional ownership and tokenization of property, starting new avenues for derivatives trading. Asset tokenization (Kondor et al., 2014) entails representing bodily assets, such as actual estate or artworks, as virtual tokens on a blockchain. Those tokens can then be traded as derivatives, imparting fractional (Double-spending, 2016) possession of traditionally illiquid assets. This innovation ought to democratize get right of entry to to formerly different asset training, developing new investment opportunities for a broader variety of investors. But, the mixing of blockchain era into the derivatives market isn't always with out its challenges and dangers. Scalability remains a subject, as the current blockchain infrastructure may additionally face barriers in handling the excessive transaction volumes generally associated with derivatives buying and selling. Moreover, regulatory uncertainties and compliance necessities call for careful consideration to make sure that blockchain-based (CoinMarketCap, 2016) totally solutions adhere to present economic guidelines and offer sturdy security features. Notwithstanding these demanding situations, the potential advantages of blockchain generation inside the derivatives market are obtrusive. As more monetary institutions and market participants include blockchain-based answers, network outcomes are probably to pressure elevated performance and liquidity, developing a virtuous cycle of adoption. The future of blockchain within the derivatives marketplace holds colossal promise, with the capability for decentralized finance (Herrera-Joancomart, 2015) (DeFi) derivatives platforms (Proof-of-Stake, 2016), in which participants can create, trade, and settle by-product contracts without intermediaries. Even as navigating regulatory concerns and technological hurdles, stakeholders within the derivatives market ought to collaborate to harness the transformative capacity of blockchain technology, fostering a extra comfortable, on hand, and modern derivatives landscape for the advantage of the global economic ecosystem.

6.2 UNDERSTANDING DERIVATIVES TRADING: A PRIMER

6.2.1 DEFINITION OF DERIVATIVES

Derivatives are economic units whose cost derives from an underlying asset, reference fee, index, event, or circumstance. They allow marketplace individuals to manipulate risk, speculate on fee moves, and reap numerous monetary objectives. While the value of a spinoff is connected to the overall performance of the underlying asset, it does no longer constitute ownership of the asset; as an alternative, it represents an agreement or agreement between two events.

6.2.2 TYPES OF DERIVATIVES IN THE FINANCIAL MARKET

1. **Futures Contracts:** These contracts entail an settlement among two events to shop for or promote an asset at a predetermined charge on a detailed future date. Commonplace underlying property include commodities like gold, oil, and agricultural products, along with monetary gadgets which include inventory indices and interest fees. Airlines, for instance, may make use of futures contracts to stabilize the rate of jet gas and mitigate operational cost risks.
2. **Options Contracts:** Alternatives offer the holder with the proper, however no longer the obligation, to buy (call option) or sell

(placed option) an underlying asset at a predetermined rate (strike rate) within a particular time body. Buyers regularly use put alternatives to hedge against capacity rate declines in stocks they maintain.

3. **Swaps:** These agreements contain parties exchanging cash flows based on precise economic variables. Hobby fee swaps and foreign money swaps are common sorts, hired for dealing with interest fee and forex risks, respectively. As an instance, an organization might enter into an interest price switch to convert variable interest bills into fixed payments.

4. **Forwards Contracts:** Similar to futures contracts, forwards agreements involve buying or selling an asset at a predetermined fee on a particular future date. But, forwards are generally customized contracts traded over-the-counter, presenting flexibility in agreement specifications. An importer would possibly use an ahead agreement to comfy a set change rate for getting foreign currency, for that reason mitigating currency fluctuation dangers (Figure 6.1).

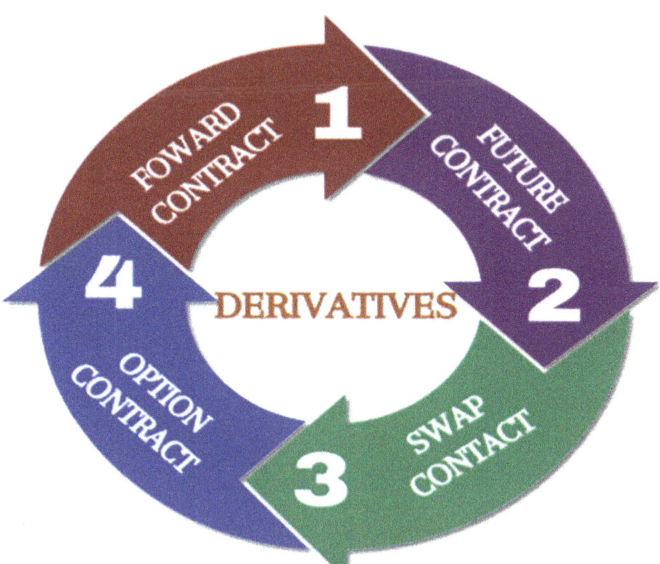

FIGURE 6.1 Types of derivatives in the financial market.

6.3 BLOCKCHAIN TECHNOLOGY: FOUNDATIONS AND FEATURES

6.3.1 DEFINITION OF BLOCKCHAIN AS A DISTRIBUTED LEDGER

Blockchain serves as a decentralized and distributed digital ledger designed to safely and transparently record, store, and verify transactions across a network of computers. It functions as an immutable and chronological chain of data blocks, each containing a set of transactions or data.

1. **Core Components of Blockchain:**

    a. **Blocks:** These containers hold a set of transactions or information, along with a unique identifier, timestamp, and reference (hash) to the preceding block, creating a tamper-evident record.

    b. **Transactions:** Representing individual movements or data entries, transactions are grouped within blocks and established for validity before being added to the blockchain.

    c. **Cryptographic Hashing:** This process converts information into precise strings of characters (hashes) for data immutability and tamper detection.

    d. **Consensus Mechanisms:** Protocols like Proof of Work (PoW) or Proof of Stake (PoS) ensure agreement on the blockchain's state among contributors.

    e. **Decentralization:** Blockchain operates on a distributed network of nodes, reducing reliance on a central authority and improving security and trust.

    f. **Immutability:** Implemented via cryptographic hashing and block interlinking, immutability ensures data integrity once added to the blockchain.

    g. **Distributed Ledger:** All transactions are distributed across collaborating nodes, ensuring transparency and redundancy without relying on a central authority.

6.3.2 TYPES OF BLOCKCHAINS

There are, in most cases, three sorts of blockchains: public blockchains, private blockchains, and consortium blockchains. Each type has distinct characteristics, use cases, and implications for privacy, security, and control. Let's delve into each kind explained in the subsections.

6.3.2.1 PUBLIC BLOCKCHAINS

Public blockchains are open and permissionless networks wherein anybody can participate, view, and confirm transactions. They're commonly decentralized and maintained by a distributed community of nodes. Public blockchains are known for their transparency, security, and censorship resistance. Some public blockchains include Bitcoin and Ethereum.

- **Advantages:**
  - Open access for everyone to participate.
  - High level of decentralization and security.
  - Enhanced transparency and immutability.
- **Challenges:**
  - Scalability and performance issues.
  - Limited privacy for participants.

6.3.2.2 PRIVATE BLOCKCHAINS

Private blockchains are restricted networks in which participation is controlled via a single entity or a set of entities. These blockchains are frequently used within companies to enhance performance, streamline processes, and maintain control over data. Private blockchains provide more privacy and control but sacrifice some of the decentralization advantages of public blockchains.

- **Advantages:**
  - Enhanced privacy and data control.
  - Improved scalability and performance.
  - More suitable for regulatory compliance.

➢ **Challenges:**

- Reduced decentralization and potential single points of failure.
- Limited censorship resistance.

6.3.2.3 CONSORTIUM (FEDERATED) BLOCKCHAINS

Consortium blockchains are a hybrid between public and private blockchains. They may be maintained through a group of pre-selected participants, often from multiple groups or entities. Consortium blockchains provide a balance between decentralization and control, making them appropriate for industries where collaboration and data sharing are critical.

➢ **Advantages:**

- Greater control compared to public blockchains.
- Enhanced privacy amongst trusted participants.
- Distributed governance and decision-making.

➢ **Challenges:**

- Governance complexities amongst consortium contributors.
- Potential reduction in decentralization in comparison to public blockchains.

6.3.3 FEATURES OF BLOCKCHAIN

1. **Decentralization:** Blockchain operates on a decentralized network of computer systems, doing away with the need for a government and decreasing single points of failure.
2. **Transparency:** Transactions on a blockchain are visible to all contributors, promoting trust and accountability.
3. **Immutability:** Data recorded on a blockchain is extremely difficult to alter once confirmed through consensus mechanisms.
4. **Security:** Blockchain employs cryptographic techniques to secure data and transactions, ensuring the integrity of the system.
5. **Privacy:** While public blockchains offer limited privacy, advancements like zero-knowledge proofs aim to enhance privacy for

sensitive information, with private and consortium blockchains providing more control over data visibility (Figure 6.2).

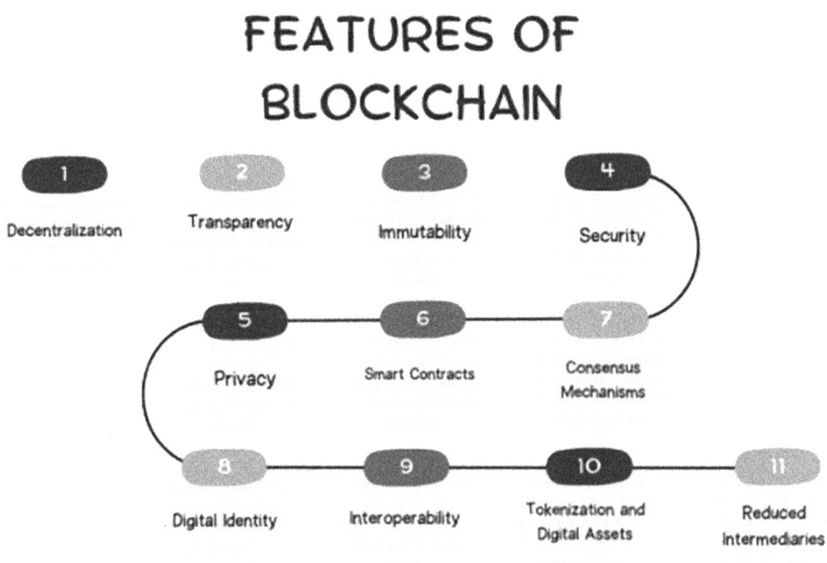

FIGURE 6.2 The features of blockchain.

6.3.4 *BLOCKCHAIN APPLICATIONS*

1. **Cryptocurrencies and Digital Belongings:** Blockchain facilitates peer-to-peer transactions without intermediaries, providing decentralized forms of cost exchange.
2. **Supply Chain Management:** Blockchain improves transparency and traceability in supply chains, reducing fraud and identifying inefficiencies.
3. **Healthcare and Medical Data:** Blockchain securely stores and manages patient medical information, improving privacy and interoperability.
4. **Identification Verification:** Blockchain provides a secure and decentralized way to manage digital identities, reducing the risk of identification theft.

5. **Voting and Governance:** Blockchain-based voting systems enhance the security and transparency of elections, providing tamper-proof records of votes.

6.3.5 CHALLENGES AND FUTURE OUTLOOK OF BLOCKCHAIN TECHNOLOGY

1. **Challenges:**
   a. **Scalability:** Public blockchains face challenges in scaling to efficiently handle increasing transaction volumes.
   b. **Energy Consumption:** PoW consensus mechanisms consume significant energy, prompting exploration of more energy-efficient alternatives.
   c. **Interoperability:** Different blockchains operate in isolation, hindering seamless communication and data exchange.
   d. **Law and Compliance:** Evolving regulatory landscapes pose challenges in balancing innovation with compliance.
   e. **Privacy and Confidentiality:** Balancing transparency with privacy remains a challenge, particularly for sensitive applications.
2. **Future Outlook:**
   a. **Mainstream Adoption:** Blockchain is moving toward broader acceptance as scalability and usability improve.
   b. **Decentralized Finance (DeFi):** DeFi is revolutionizing traditional financial services by offering decentralized lending, trading, and yield-generating systems.
   c. **Non-Fungible Tokens (NFTs):** NFTs are tokenizing and authenticating digital art, collectibles, and more, creating new revenue streams for creators.
   d. **Corporate Integration:** Corporations are recognizing blockchain's benefits for supply chain management, data sharing, and transparency.

e. **Central Bank Digital Currencies (CBDCs):** Nations are exploring CBDCs, which could transform money supply and transaction management.

6.4 DISRUPTION OF TRADITIONAL DERIVATIVES TRADING: NAVIGATING THE NEW LANDSCAPE

6.4.1 DERIVATIVES: DEFINITION AND ROLE IN HAZARD MANAGEMENT

Derivatives constitute monetary instruments whose fee derives from an underlying asset, index, or reference price. These contracts among events outline situations for exchanging payments based totally at the future fee of the underlying asset. Derivatives, without intrinsic price, rely upon the charge movements of the underlying asset for valuation.

6.4.2 KINDS OF DERIVATIVES

1. **Alternatives:** Contracts granting the holder the proper, but now not the duty, to buy (name option) or promote (placed option) an underlying asset at a predetermined fee (strike charge) inside a unique time frame. Alternatives are normally used for hedging towards market fluctuations and for speculation.
2. **Futures Contracts:** Standardized agreements to buy or promote an underlying asset at a predetermined rate on a selected future date. Not like options, futures contracts bind both events to fulfill the settlement. They are often utilized for hedging against charge volatility and for publicity to commodities, currencies, and other belongings.
3. **Swaps:** Agreements among parties to trade cash flows based on certain terms. Commonplace types include interest fee swaps, foreign money swaps, and commodity swaps. Swaps are essential in coping with risks associated with hobby rate modifications, foreign money fluctuations, and commodity price movements.

4. **Forward Contracts:** Agreements to shop for or promote an underlying asset at a future date and a negotiated fee. Not like futures contracts, ahead contracts are customizable and no longer traded on exchanges. They discover utility in hedging unique exposures, including commodities or currencies.

This classification encompasses the primary styles of derivatives, every serving distinct functions and addressing particular chance control wishes.

6.5 CONCLUSION

In summary, consortiums play an essential role in cross-border derivatives settlement, main to modernization and optimization within the worldwide monetary panorama. Despite dealing with challenges consisting of regulatory complexity, technological integration, and threat management, it's critical to apprehend the transformative potential of blockchain technology in this dynamic area. Delving deeper, we discover how blockchain serves as a cornerstone in enhancing efficiency, decreasing fees, improving danger control, and fostering elevated marketplace liquidity and transparency.

One of the modern elements of blockchain technology is its ability to provide a at ease, obvious, and decentralized ledger for financial transactions. The allotted nature guarantees statistics integrity across a community of computers, decreasing the threat of fraud or manipulation. Inside the context of move-border derivatives agreement, in which believe and safety are paramount, blockchain emerges as a powerful solution to beautify the integrity of economic operations. Efficiency is another cornerstone of blockchain era, streamlining go-border derivatives agreement processes through clever contracts. These self-executing contracts automate various stages of settlement, decreasing the need for intermediaries and minimizing errors and delays. The time-touchy nature of financial transactions is well-addressed with the aid of blockchain, contributing to more advantageous operational efficiency. Price reduction is a compelling benefit of blockchain, allowing direct peer-to-peer transactions and minimizing related prices thru automation facilitated through clever contracts.

Blockchain's effect on danger management is great, with immutable facts making sure transparent transaction histories and mitigating the chance of fraud or disputes. Real-time visibility into transactions enhances the ability of financial establishments to deal with capability risks right away. Moreover, blockchain contributes to elevated marketplace liquidity and transparency, fostering consider among contributors and attracting extra marketplace players.

In essence, integrating blockchain technology within consortiums engaged in move-border derivatives agreement represents a tremendous soar inside the evolution of the worldwide financial panorama. Leveraging blockchain's abilities addresses present challenges while unlocking new possibilities for performance, price reduction, risk mitigation, and market transparency. As consortiums maintain to champion collaboration, standardization, and innovation, their adoption of blockchain technology solidifies their position as trailblazers using nice transformation inside the financial industry.

KEYWORDS

- **asset tokenization**
- **clever contracts**
- **cryptocurrencies**
- **derivatives**

REFERENCES

Ali, S., McCorry, P., Lee, P. J., & Hao, F. (2015). ZombieCoin: Powering next-generation botnets with Bitcoin. In M. Brenner, N. Christin, B. Johnson, & K. Rohloff (Eds.), *Financial cryptography and data security* (Vol. 8976, pp. 34–48). Springer Berlin Heidelberg. http://dx.doi.org/10.1007/978-3-662-48051-9_3 (accessed on 22nd September 2025).

Andrychowicz, M., Dziembowski, S., Malinowski, D., & Mazurek, M. (2015). On the malleability of Bitcoin transactions. In M. Brenner, N. Christin, B. Johnson, & K. Rohloff (Eds.), *Financial cryptography and data security* (Vol. 8976, pp. 1–18). Springer Berlin Heidelberg. http://dx.doi.org/10.1007/978-3-662-48051-9_1 (accessed on 22nd September 2025).

Antonopoulos, A. M. (2014). *Mastering Bitcoin: Unlocking digital cryptocurrencies*. O'Reilly Media, Inc.

Ateniese, G., Faonio, A., Magri, B., & de Medeiros, B. (2014). Certified Bitcoins. In I. Boureanu, P. Owesarski, & S. Vaudenay (Eds.), *Applied cryptography and network security* (Vol. 8479, pp. 80–96). Springer International Publishing. http://dx.doi.org/10.1007/978-3-319-07536-5_6 (accessed on 22nd September 2025).

Bitcoincharts. (2016). https://bitcoincharts.com (accessed on 22nd September 2025).

Bitcoinwiki. (2015). https://en.bitcoin.it (accessed on 22nd September 2025).

Business Electricity Prices. (2017). A guide to wholesale electricity pricing. https://www.businesselectricityprices.org.uk/retail-versus-wholesale-prices/ (accessed on 22nd September 2025).

Coinmarketcap. (2016). Crypto-currency market capitalizations. https://coinmarketcap.com/ (accessed on 22nd September 2025).

Decker, C., & Wattenhofer, R. (2014). Bitcoin transaction malleability and MtGox. In M. Kutyłowski & J. Vaidya (Eds.), *Computer security—ESORICS 2014* (Vol. 8713, pp. 313–326). Springer International Publishing. http://dx.doi.org/10.1007/978-3-319-11212-1_18 (accessed on 22nd September 2025).

Decker, C., Guthrie, J., Seidel, J., & Wattenhofer, R. (2015). Making Bitcoin exchanges transparent. In G. Pernul, P. Y. A. Ryan, & E. Weippl (Eds.), *Computer security—ESORICS 2015* (Vol. 9327, pp. 561–576). Springer International Publishing. http://dx.doi.org/10.1007/978-3-319-24177-7_28 (accessed on 22nd September 2025).

Dev, A. J. (2014). Bitcoin mining acceleration and performance quantification. In *Electrical and Computer Engineering (CCECE), 2014 IEEE 27th Canadian Conference on* (pp. 1–6).

Double-spending. (2016). https://en.bitcoin.it/wiki/Double-spending (accessed on 22nd September 2025).

Dybå, T., & Dingsøyr, T. (2008). Empirical studies of agile software development: A systematic review. *Information and Software Technology, 50*(9–10), 833–859. http://dx.doi.org/10.1016/j.infsof.2008.01.006 (accessed on 22nd September 2025).

Herrera-Joancomartí, J. (2015). Research and challenges on Bitcoin anonymity. In J. Garcia-Alfaro, J. Herrera-Joancomartí, E. Lupu, J. Posegga, A. Aldini, F. Martinelli, et al. (Eds.), *Data privacy management, autonomous spontaneous security, and security assurance* (Vol. 8872, pp. 3–16). Springer International Publishing. http://dx.doi.org/10.1007/978-3-319-17016-9_1 (accessed on 22nd September 2025).

Housley, R. (2004). In *Public key infrastructure (PKI)*. John Wiley & Sons, Inc. http://dx.doi.org/10.1002/047148296X.tie149 (accessed on 22nd September 2025).

Kitchenham, B., & Charters, S. (2007). *Guidelines for performing systematic literature reviews in software engineering*.

Kondor, D., Pósfai, M., Csabai, I., & Vattay, G. (2014). Do the rich get richer? An empirical analysis of the Bitcoin transaction network. *PLOS ONE, 9*(2), e86197.

Nakamoto, S. (2008). Bitcoin: A peer-to-peer electronic cash system. *Consulted, 1*(2012), 28.

Proof-of-Stake. (2016). https://en.bitcoin.it/wiki/Proof_of_Stake (accessed on 22nd September 2025).

Swan, M. (2015). *Blockchain: Blueprint for a new economy*. O'Reilly Media, Inc.

CHAPTER 7

Blockchain Techniques for Fraud Prevention in the Financial Market: A Comprehensive Review

SUNIL MUNDE[1] and ANJALI MUNDE[2,*]

[1]*Engineering Management - Topside Engineering Coordination, Kuala Lumpur, Malaysia*

[2]*Southampton Malaysia Business School, University of Southampton Malaysia, Iskandar Puteri, Johor, Malaysia*

*Corresponding author

ABSTRACT

Fraudulent actions present considerable issues in the financial system, resulting in cash losses, reputational damage, and lost client trust. However, blockchain technology, because to its decentralized and immutable nature, presents intriguing options for preventing financial fraud. This chapter provides a comprehensive overview of blockchain strategies used to fight fraud in the financial markets. The first section of the chapter presents an overview of financial market fraud, stressing its various varieties, effect, and challenges to existing mechanisms. The second portion delves into the fundamentals of blockchain technology, including its major qualities such as decentralization, transparency, and immutability. The final segment goes into specific blockchain strategies used for fraud prevention. It covers smart contracts and their potential to automate enforcement and improve security in financial transactions. The fourth section discusses the problems and limitations of applying blockchain solutions for fraud prevention. The fifth section discusses prominent use cases and success stories

FinTech and Blockchain Technology in Financial Markets. Neha Puri, Anjali Munde, Emilia Alaverdov (Eds.)
© 2027 Apple Academic Press, Inc. Co-published with CRC Press (Taylor & Francis)

DOI: 10.1201/9781779641298-7

for blockchain-based fraud prevention solutions in the financial business. Finally, the chapter discusses blockchain technology's future prospects for preventing financial fraud. Through this comprehensive study, the chapter intends to provide scholars, practitioners, and policymakers words with a clearer knowledge of the potential of blockchain technology in combating fraud in the financial industry.

7.1 OVERVIEW OF FRAUD IN THE FINANCIAL MARKET

Fraud, according to Transparency International (TI), is the misuse of (public) power for one's own gain (Vargas-Hernández, 2009). Even in countries with strict laws and conservative cultures, fraud exists. Organized crime, fraud, and terrorism can pose significant risks to the financial industry. The financial industry undoubtedly plays an important influence in shaping how a society evolves (Goldsmith, 1969; Levine, 2003; Rajan and Zingales, 1998). One of the main impediments to the rule of law and respect for human rights in the financial industry is fraud. The whole financial sector, including financial markets, international finances, state finances, business finances, and personal finances, are now concerned about it. The most enduring impediment to every country's ability to progress economically and socially is fraud (Keefer and Knack, 1997; Li et al., 2000; Méon and Sekkat, 2005; Mo, 2001). Additionally, fraud is linked to the growth of organised crime (Hardouin, 2009). According to Gyimah-Brempong (2002), fraud has a significant negative impact on the economic growth and income distribution of some nations, particularly those with a lower level of democracy. The disruption of competition, the impediment of both domestic and international investments, and the slowing of growth are all effects of corruption in the financial ecosystem. Other firms that operate and thrive in accordance with the law are hampered by such acts. Fraud and organised crime drive people away. Due to widespread employee unhappiness, distrust, and job insecurity in emerging economies, the situation is more concerning (Aziz et al., 2007; Lim et al., 2023). According to Venard (2008), financial institution fraud reduces the confidence of the institutions and the market both domestically and abroad. It causes a reduction in public trust in the organization and could harm the business. Since fraud is a broad topic, academic researchers in a variety of fields, including finance, study it. Even though academic research in this field started in the 1980s, it really took off in the 1990s.

Academic scholars' interests coincided with the economic and financial ecosystem's engagement with globalization, information technology, and the internet, resulting in the creation of a substantial body of literature. The study, however, is dispersed, with publications addressing diverse subjects in distinct contexts. Furthermore, it appears that no study has yet been done that offers a comprehensive analysis or discussion of the problem of financial fraud. There aren't many literature studies on this subject that talk about particular circumstances. For example, Khelil (2021) examines the empirical data on the economic implications of political affiliations and political corruption on auditor behavior in the accounting and finance fields in their article from 2021. In their review of the literature, Schnatterly (2018) provide insights into the causes of CEO misbehavior. This makes it appropriate to conduct a critical investigation of the area as a whole, and the purpose of this study is to give a full and in-depth analysis of the current state of corruption in finance research. As Lim (2022) pointed out, it's equally crucial to make a strong case for closing the gap that's now there. Financial fraud is a major threat in the corporate sector. Despite technological advancements, fraud cases are increasing (Beena et al., 2021). According to one study Online Payment Fraud Losses to Exceed $206 Billion Over the Next Five Years; Driven by Identity Fraud (2022), between 2021 and 2025, online transaction theft is expected to cost retailers $206 billion in lost revenue. Authentic data is required for the development of reliable fraud detection algorithmic models. However, financial transaction data is rarely available for machine learning research due to its private nature, privacy concerns, and a lack of organizational synergy (Amarasinghe et al., 2018; Assefa et al., 2020). Furthermore, organizations usually avoid performing collaborative study on financial issues in order to conceal their financial reports and business goals (Amarasinghe et al., 2018). The majority of systems and services globally are owned by large corporations, who also handle massive amounts of data (Rikap and Lundvall, 2021). Crowdsourcing is a common approach for researchers from small firms to collect data; however, it is prone to spoofing and frequently produces completely meaningless data (Lu et al., 2020). Fraud in the financial market refers to deceptive or illegal practices conducted within the realm of financial transactions and activities. It involves deliberate acts of dishonesty with the intention of obtaining financial gain or causing financial harm to individuals, organizations, or the economy as a whole. Financial market fraud can occur across various sectors, including banking, investment, insurance, and securities trading. Understanding the various types, consequences, and obstacles connected with financial market fraud is critical for devising effective ways to resist and prevent such illegal activity.

7.1.1 TYPES OF FINANCIAL MARKET FRAUD

Identity theft is the illicit acquisition and use of another person's personal information, such as social security numbers, credit card numbers, or passwords, in order to conduct fraudulent financial transactions. This includes setting up false accounts, doing unlawful activities, and acquiring loans or credit in the victim's name. Transactional fraud refers to a variety of fraudulent operations aimed at manipulating financial transactions. This can include faking documents, fabricating invoices, manipulating payment systems, or engaging in unlawful trades. Transactional fraud frequently leverages flaws in transaction processes and systems to acquire a financial advantage or deceive stakeholders. Market manipulation refers to deliberate actions that alter the natural operation of financial markets for personal advantage. This can include spreading false information to influence stock prices, engaging in insider trading based on non-public information, carrying out pump-and-dump schemes to inflate stock prices, or engaging in front-running, in which traders use advance knowledge of impending transactions to their benefit. Ponzi schemes are fraudulent investment schemes that promise high returns to early investors using funds from subsequent investors. The scheme collapses when there are no more new investors, resulting in significant financial losses for those involved.

Money laundering is the process of making illegally obtained funds appear legitimate by disguising their origins. It entails a sequence of intricate transactions that conceal the origin of the funds and integrate them into the lawful financial system, making it difficult to identify and uncover illegal operations.

7.2 INTRODUCTION TO BLOCKCHAIN TECHNOLOGY

Satoshi Nakamoto first outlined the concepts of Bitcoin and blockchain in 2008, describing how encryption and an open distributed ledger may be used to build a digital currency application (Nakamoto, 2008). Bitcoin's growth was initially impeded by its great volatility and the viewpoints of many countries on its complexity, but the benefits of blockchain—the technology that supports Bitcoin—attracted increasing interest. Blockchain benefits include distributed ledgers, decentralization, information transparency, tamper-proof design, and openness. Blockchain development progressed over time. Blockchain technology is now divided into three versions based on the applications it can support: 1.0, 2.0, and 3.0. In the Appendix, we provide more information on the three blockchain generations. Blockchain technology is increasingly used for more than only digital currency and finance, but also for market monitoring, supply chain management, copyright protection, and smart energy (Engelhardt, 2017; Hyvarinen et al., 2017; Kim and Laskowski, 2018; O'Dair and Beaven, 2017; Radanovic and Likic, 2018; Savelyev, 2018).

Many academic fields have investigated blockchain technology. For example, several researchers have looked into the blockchain's underlying technologies, such as distributed storage, peer-to-peer networking, cryptography, smart contracts, and consensus methods (Christidis and Devetsikiotis, 2016; Cruz et al., 2018; Kraft, 2016). Legal researchers are interested in the rules and laws that regulate blockchain technology (Kiviat, 2015; Paech, 2017). According to an old saying, scholars from various fields "speak many different languages" and have a wide range of analytical perspectives.

7.2.1 BLOCKCHAIN

A distributed ledger system known as a blockchain allows all network participants to share transaction information between the parties. In other words, a blockchain is a distributed database that keeps track of an ever-expanding list of data entries that are protected against alteration and revision, even by the people who run the nodes that make up the data store. Blocks, which are collections of transactions that happened over a 10-minute period and are theoretically connected to form a blockchain, are used to validate transaction information. The blockchain is a public ledger that is accessible to everyone on the network and contains a chronological record of all previous Bitcoin transactions. In order to ensure the dependability of transaction records without being informed by a "trusted third party" (TTP), such as a central bank or administrative agency, all network participants record, validate, and preserve transaction information collectively. Because it updates every member's ledger whenever a new transaction takes place, this system is also incredibly safe. The blockchain cannot be falsified or compromised as long as more than 51% of participants are in sync and write transaction information on the distributed ledger. When paired with a public key algorithm, a hash encryption method, and a distributed processing structure, blockchain technology can have a low cost impact. As a result, a distributed ledger may replace the current centralized ledger system. The banking industry sees advantages in terms of stability, security, and reduced costs for data maintenance. As a result, service delivery times are likely to speed up. The pay-per-view transportation system is particularly vulnerable to blockchain technology since peer-to-peer financial transactions between players may be completed with only an internet connection, eliminating the need for a banking institution or TTP. Transaction transparency promotes information security, and we are exploring the use of blockchains in a range of industries, including finance, health, and administration. As a result, start-ups are rapidly extending blockchain-based services in settlement, remittance and lending, banking, securities, and investing.

7.2.2 BLOCKCHAINS IN THE FINANCIAL SECTOR

Modern electronic payment systems rely on centralized, reputable third parties to ensure payment security. The need to reduce transaction costs led banks to begin accepting claims from one another. Traders no longer had to worry about converting paper money to gold when transferring currency; instead, they could deposit notes from other banks directly into their own, making business easier. However, by accepting the note from a different bank, the payee's bank created a new problem for itself, as it was now exposed to the payer until a gold payment agreement could be made. When only a few banks accept notes, a bilateral solution may be used. Interbank payments became increasingly sophisticated as the number of banks in the system increased, putting pressure on banks to find more efficient alternatives. Bitcoin, a digital currency that mixes new currencies with decentralized payment networks, is still relatively new. Although the financial aspects of digital currencies have garnered a lot of attention, the distributed ledger that powers their payment systems marks a huge step forward. Today, the majority of financial assets exist solely as digital records, similar to money held in bank accounts. Distributed ledger technology has the ability to transform the financial industry as a whole. The distributed ledger, often known as a blockchain, can be classified as closed and private, closed and public, or open and public based on participation and ownership guidelines. If it is not permitted, it is open to all. "Permitted" refers to a construction that only permitted people can enter. A private ledger is also limited to a small number of interested parties, whereas a public ledger is accessible to everybody. We use a closed distributed ledger to process interbank payments and international financial transactions. Blockchains with a closed distributed ledger that only authorized users may access are preferred because these characteristics are critical in the financial industry. To compensate for the drawbacks of openness, the closed type includes a consensus method that certifies transaction validity while restricting participation to a small number of specialized organizations. The key objectives are to ensure technical advancement and homogenization. Because of the lack of new standard techniques brought about by technology innovation, open type is difficult to standardize; nevertheless, closed type participants can quickly agree on and approve technical standards. Second, it can operate independently and successfully. The open type decreases specialized power and the need for agency communication, which is beneficial; yet, it has a poorer efficiency structure in consensus structures than the closed type. Third, the transaction can be changed in the closed type case. A closed type transaction can be altered with mutual approval; however, an open type transaction cannot be amended once entered into the spreadsheet and must be repaired using reverse trading. The financial industry uses a closed distributed ledger in this regard.

7.2.3 PREVIOUS RESEARCH

Blockchain has recently become a focus of research in a variety of businesses. According to Lee and Kim (2016), to address numerous security concerns, we provide a low-cost and secure smart grid system authentication technique based on blockchain, a key component of Bitcoin, and demonstrate how it may be utilized to authenticate system components. According to Hwa (2016), in order to employ distributed ledger technology, the current centralized regulatory system must be overhauled. There are several legal issues to address, including the location of physical data storage, the legal foundation for regulatory body action, and the blockchain's standard protocol and governance. Because public blockchain deployment is limited, financial institutions are more likely to adopt blockchain technology via a consortium or a private blockchain. Marc Pilkington's 2015 paper explores the core concepts of blockchain, as well as the benefits and drawbacks of public distributed ledger technology and the shift to hybrid solutions. He covered the main components of platforms that support decentralized public ledgers. Blockchain technology is a decentralized, distributed ledger system that allows for secure and transparent transactions without the use of intermediaries. It was originally meant to be the foundational technology underpinning the cryptocurrency Bitcoin, but it has since found applications in a variety of fields. Blockchain is based on the creation and maintenance of an ever-expanding chain of blocks, each carrying a series of confirmed and time-stamped transactions. A blockchain is a decentralized network in which individual members, known as nodes, store copies of the entire blockchain. Decentralization eliminates the need for a central authority, such as a bank or government, to oversee and approve transactions. Blockchain fosters transparency by making all transactions available to network members. This transparency encourages accountability by allowing users to audit and validate changes to the blockchain. Once a transaction has been recorded on the blockchain and included in a block, it is nearly impossible to modify or tamper with. The blockchain's immutability ensures that earlier data is retained and secured, resulting in an auditable and tamper-proof transaction log. Blockchain networks employ consensus procedures to obtain an agreement among participating nodes about the legality and sequence of transactions. The three most popular consensus mechanisms are Proof of Work (PoW), Proof of Stake (PoS), and Practical Byzantine Fault Tolerance (PBFT). These protocols ensure that all network nodes are in agreement on the blockchain's present status.

7.2.4 BLOCKCHAIN COMPONENTS

Blockchain technology is made up of several critical components that work together to build a decentralized, secure system. Blocks are the fundamental units of a blockchain. Each block comprises a set of transactions and other pertinent information. It contains a unique identification known as a cryptographic hash, which ensures the block's integrity and contents. The hash of each block includes the hash of the previous block, resulting in a chain of blocks. Transactions are the exchange of data, assets, or value across the blockchain network. They can contain a variety of information, including financial transactions, contracts, and asset transfers. Each transaction is digitally signed with cryptographic techniques to ensure its validity and integrity. Cryptographic hashing is a mathematical procedure for converting any input data into a fixed-length string of letters known as a hash. This hash is unique to the input data and functions as a digital fingerprint. Hashes are employed in blockchains to maintain data integrity and securely connect blocks. A distributed ledger is a decentralized database that replicates the whole blockchain across multiple nodes in the network. Each node has its own copy of the ledger, and they collaborate to validate and save transactions. This distributed architecture ensures that the system is transparent, secure, and redundant. Consensus mechanisms are protocols or algorithms that allow nodes in a blockchain network to agree on the authenticity and ordering of transactions. They ensure that all parties agree on the status of the blockchain. PoW, PoS, and PBFT are among the most popular consensus procedures. Wallets are software apps or hardware devices that allow participants to safely keep their cryptographic keys. These keys are required to access and manage the assets or data linked with a given blockchain address. Wallets allow users to transmit and receive transactions over the blockchain. A peer-to-peer network is the architecture that connects the nodes in a blockchain system. It makes it easier for nodes to communicate and share information, ensuring the blockchain's decentralization. Each node maintains a copy of the blockchain and participates in transaction validation and consensus. Smart contracts are self-executing contracts with predefined rules and conditions encoded on the blockchain. They automate and enforce the terms of an agreement, eliminating the need for intermediaries. Smart contracts execute actions automatically when certain conditions are met, enabling trust and efficiency in various applications. These components work together to build a transparent, secure, and decentralized system that serves as the foundation for blockchain technology. Each component is critical to maintaining the blockchain network's integrity, immutability, and reliability.

7.3 BLOCKCHAIN TECHNIQUES FOR FRAUD PREVENTION

Chao Yang's research presents an incremental outlier clustering method based on blockchain that is more performant and computationally efficient (Yang, 2022). Machine learning techniques are used for their predictive capabilities, and blockchain technology is rapidly gaining traction to increase system security (Shahbazi and Byun, 2021). Following Bitcoin's phenomenal success, the scientific and corporate communities have shifted their attention to the breakthrough technology known as blockchain. Blockchain is a distributed ledger technology that preserves transaction and data logs using encryption (Niranjanamurthy et al., 2019). A peer-to-peer network stores transaction or data records as information blocks (Nofer et al., 2017). Each block is linked to the previous one, and the genesis block serves as the beginning point (Haque et al., 2021). Blockchain's primary purpose is to keep transaction records decentralized, immutable, public, accessible, and safe while preserving anonymity (Nofer et al., 2017). Blockchain is a decentralized peer-to-peer network that is not managed by a single authority.

No one node or group of nodes has complete control over the data; rather, all connected nodes have equal access to it (Chen and Bellavitis, 2020). Immutability, or the inability to change data, is an important feature of the blockchain. Blockchain maintains an increment-only digital record; once data is published to the network, it cannot be changed or deleted (Hofmann et al., 2017). Every node connected to a blockchain stores a copy of the most recent ledger (Niranjanamurthy et al., 2019). All parties involved in the blockchain network have access to the data kept there, and because this data is always available, the system is open and accessible (Rizal Batubara et al., 2019). In addition to these features, blockchain is marketed as being incredibly safe. Because of its distinct characteristics, blockchain has been utilized to address a wide range of faults in older systems (Haque et al., 2021; Zheng et al., 2019). Blockchain technology has been successfully implemented in the financial industry (Varma, 2019). Blockchain eliminates double spending by storing timestamps for each transaction under cryptographic security, as well as transaction identification (Iqbal and Matulevičius, 2021). Furthermore, the distributed consensus process ensures that the majority of nodes accept the new transaction (Nofer et al., 2017). Iqbal and Matulevicius examined these issues further and devised a method for detecting double-spending early on (Iqbal and Matulevičius, 2021). Blockchain-based solutions offer extremely safe and reliable media for commercial and trade applications such as insurance, investment transactions, venture capital financing, and so on (Varma, 2019). For example, Roriz and Pereira created a blockchain-based solution for double-dipping (having numerous insurance policies on the same vehicle), a sort of motor insurance fraud (Roriz and Pereira, 2019). This technology has also been used in healthcare, education, process automation, and supply chain management, as well as finance (Bhaskar et al., 2020; McGhin et al., 2019). Blockchain has many benefits, but it requires additional skills to display complicated transactions when used in real-world applications (Nofer et al., 2017). Blockchain Techniques for Fraud Prevention take advantage of the technology's transparency, immutability, decentralized identity, smart contracts, and real-time monitoring capabilities to enhance security, identify fraud, and maintain trust in financial markets. One of the key features of blockchain technology is the creation of a transparent and immutable ledger. Blockchain maintains a transparent record of all transactions that is available to all network participants. This transparency enables real-time monitoring and auditing, making it impossible to hide fraudulent activities. Once a transaction is recorded on the blockchain, it is nearly impossible to alter or tamper with. Blockchain's immutability ensures the integrity and authenticity of transaction data, making it an important tool for fraud prevention.

Blockchain enables transaction traceability, allowing auditors and investigators to monitor the flow of funds and identify suspect activity. This auditability promotes responsibility and serves as a disincentive for potential scammers. Furthermore, identity theft is a widespread type of financial crime. Blockchain technology provides numerous possibilities for preventing and mitigating identity theft. Blockchain technology can provide decentralized and self-governing identification solutions. Users can retain control of their identification information kept on the blockchain, lowering the danger of identity theft by minimizing reliance on centralized systems susceptible to hacking. Blockchain technology enables safe, tamper-proof identity verification processes. Organizations that store validated identification information on the blockchain can validate user IDs without using traditional centralized identity verification methods. Blockchain technology enables multi-factor authentication systems, which require many parties to confirm a transaction or identity before it is considered valid. This provides an extra layer of security and lowers the chance of identity theft. Blockchain technology can be used to prevent and detect fraudulent financial transactions. Smart contracts are blockchain-based agreements that run automatically according to predefined criteria and conditions. They can automate and enforce transactional operations, ensuring that transactions occur only when particular conditions are met. Smart contracts help to reduce fraud by eliminating middlemen and minimizing the possibility of human error or manipulation. The immutability of blockchain ensures that once recorded, a transaction cannot be changed or tampered with. This feature makes it more difficult for fraudsters to change transaction information or create forged records. Blockchain provides real-time transaction monitoring, which detects suspicious or fraudulent activity as it happens. Blockchain data can be analyzed using advanced analytics and algorithms to detect trends that indicate fraudulent behavior. Sector manipulation is a major issue in the financial sector. Blockchain technology can help to prevent and detect market manipulation. Blockchain's transparency allows regulators and market players to track transactions and market activity in real time. This transparency lowers the danger of market manipulation while also ensuring fair and equal access to information. Market data, announcements, and press releases may all be timestamped and documented using blockchain technology. This gives proof of existence while also preventing timestamp manipulation for fraudulent purposes. Blockchain-based smart contracts can be set up to autonomously enforce regulatory compliance criteria. This decreases the likelihood of market manipulation by ensuring that participants follow regulatory guidelines. Blockchain data, when paired with advanced analytics and artificial intelligence, has the ability to drive complex market surveillance systems. These systems can detect trends and anomalies that indicate market manipulation, offering an early warning and rapid response.

7.4 CHALLENGES AND LIMITATIONS WITH IMPLEMENTING BLOCKCHAIN TECHNIQUES FOR FRAUD PREVENTION

While blockchain technology offers intriguing answers for preventing financial fraud, its implementation confronts a number of problems. Understanding these problems is critical for successfully using blockchain technology for fraud prevention. Blockchain scalability refers to a blockchain network's ability to efficiently process a high number of transactions. As the volume of transactions grows, the blockchain network may encounter problems such as slower transaction processing times and greater resource demands. These scaling issues can impede real-time transaction monitoring and responsiveness, potentially allowing fraudulent acts to go undetected or unreported. Integrating blockchain technology with existing legacy systems can be complicated and difficult. Many financial organizations use traditional centralized systems, which may be incompatible with blockchain's decentralized and distributed nature. Bridging the gap between current systems and blockchain infrastructure requires precise planning, coordination, and technology upgrades, which can be time-consuming and costly. The legislative and regulatory landscape surrounding blockchain technology is constantly evolving. Financial firms must navigate complex regulatory frameworks to meet anti-money laundering (AML) and know-your-customer (KYC) rules. Furthermore, the immutability and pseudonymous nature of blockchain can present issues in terms of data privacy, data protection, and the ability to identify and prosecute fraudsters. Addressing these legal and regulatory concerns is critical to the wider deployment of blockchain for fraud prevention. The lack of established protocols and compatibility between different blockchain systems and networks poses a significant barrier. Financial institutions, regulators, and other stakeholders may use separate blockchain platforms, making it difficult to communicate data and work successfully. The creation of industry-wide standards and protocols is critical to ensuring seamless interoperability and facilitating information exchange for fraud prevention. Blockchain networks use consensus mechanisms to determine the legitimacy and sequence of transactions. However, determining and implementing the optimum consensus procedure for fraud prevention can be challenging. Different consensus algorithms have unique trade-offs in terms of scalability, security, energy efficiency, and decision-making procedures. Choosing the right consensus technique and developing efficient governance structures within the blockchain network are crucial for combating fraud. Blockchain technology is still in its early phases, and most users and stakeholders are ignorant of it.

It is vital to promote user acceptance and educate stakeholders on blockchain technology's benefits, functionality, and security. Without enough user acceptance and education, blockchain's potential for fraud prevention may not be fully realized. Implementing blockchain technology for fraud prevention incurs significant expenditures and resource needs. Infrastructure expenditures include the installation and maintenance of blockchain nodes, as well as network security and data storage management. Furthermore, educating employees and getting the requisite technical competence to work with blockchain technology can be expensive. To achieve successful implementation, financial institutions must carefully examine the cost-benefit ratio and spend resources efficiently.

While blockchain technology has significant benefits for preventing fraud in the financial system, it does have some limits that must be recognized. Understanding these constraints is critical for making an accurate assessment of blockchain's effectiveness in combatting fraud. The inherent transparency of blockchain presents issues in terms of privacy and confidentiality. Transaction data are encrypted, but the transaction history and related wallet addresses are exposed to all network members. This lack of privacy can be problematic when dealing with sensitive financial data because it exposes parties to potential threats such as identity theft or targeted assaults. To provide effective security of confidential information, a balance between transparency and privacy must be maintained.

Blockchain networks need all participants to keep a copy of the full blockchain, which might result in large data storage requirements. As the blockchain increases in size and transactions, it becomes more difficult to maintain and manage the storage infrastructure. This issue can limit the scalability of blockchain networks and their capacity to process huge volumes of transactions efficiently. Certain blockchain consensus algorithms, such as PoW, necessitate considerable processing power and energy usage. The energy-intensive nature of these systems has led to concerns regarding blockchain technology's environmental impact. Furthermore, the high energy requirements of blockchain may make it less accessible and cost-effective for small-scale users, restricting its use for fraud prevention in some situations.

The decentralized nature of blockchain can affect transaction processing speeds. In public blockchain networks, where numerous nodes must reach consensus, validating and confirming transactions might take longer than in centralized systems. This constraint could impair real-time fraud identification and prevention, particularly in high-frequency trading or time-sensitive financial transactions.

Blockchain technology is global and decentralized, creating issues for regulatory supervision and compliance. Existing legislation and legal frameworks may not be adequate to accommodate the distinct characteristics of blockchain networks. Implementing effective regulatory measures to prevent fraud while maintaining compliance with existing rules is a difficult task. Harmonizing regulations across jurisdictions and adapting them to blockchain technology pose continual hurdles.

Building and maintaining a blockchain infrastructure can be expensive and complicated. The initial setup costs, which include hardware, software, and network infrastructure, can be high. Furthermore, the continuing maintenance and technical knowledge necessary to administer and improve the blockchain network increase the overall expenses. These issues may make it difficult for smaller firms with limited resources to implement blockchain for fraud prevention.

While blockchain technology is highly secure and immutable, it does not eliminate the possibility of human error or social engineering assaults. Fraudsters may use vulnerabilities in human behavior, like as phishing attacks or social engineering tactics, to gain unauthorized access to blockchain wallets or trick people into making fraudulent transactions. Blockchain technology alone cannot reduce these human-related hazards; a more comprehensive strategy, including user education and awareness, is needed.

Understanding these limits aids in identifying prospective situations where blockchain techniques for fraud protection may not be the best fit. It is critical to evaluate the financial market's specific requirements, restrictions, and risk profiles in order to develop the most effective combination of technologies, processes, and protections for preventing fraud.

7.5 BLOCKCHAIN-BASED FRAUD PREVENTION SOLUTIONS IN THE FINANCIAL MARKET

Blockchain-based fraud prevention solutions have gained interest in the financial business due to their ability to improve transaction security, transparency, and trust. KYC and AML procedures are critical for combating fraudulent activity in the financial sector. Blockchain technology helps improve KYC/AML compliance by securely storing client identity information and transaction data. Solutions such as SelfKey and Jumio use blockchain to offer self-sovereign identity management, ensuring safe and tamper-proof verification processes. Trade financing involves a number of partners, including importers, exporters, banks, and insurance. Blockchain-based solutions, such as we.trade and TradeLens, maintain transparent and immutable records of trade-related data, such as purchase orders, invoices, and shipping papers. This helps to avoid fraud by lowering the chances of document tampering, duplicate financing, and counterfeit items. Blockchain technology can help to streamline and automate insurance claims processes, eliminating false claims and increasing efficiency. Companies such as Insurwave and AXA use blockchain to create immutable records of policyholder data, claim history, and settlements. These technologies allow for real-time verification, reduced paperwork, and increased trust between insurers and policyholders. Blockchain-based payment solutions provide enhanced security and fraud prevention. Ripple, for example, uses blockchain technology to enable real-time, secure cross-border transactions. Its network reviews transactions and verifies participants, reducing the risk of payment fraud, such as unlawful transactions and identity theft. Blockchain-based digital identification solutions improve fraud prevention by enabling safe and verifiable identity management. Sovrin and uPort are examples of blockchain-based identification services that provide users control over their personal data while providing secure and trustworthy verification processes. These technologies decrease the likelihood of identity theft and unauthorized access to financial accounts. Blockchain technology has the ability to transform securities and trade by providing transparent and tamper-proof record-keeping. Nasdaq, for example, has adopted blockchain technology for private market transactions, allowing for more efficient and secure share issuance and transfer. This decreases the likelihood of fraudulent activity, such as unlawful share transfers and counterfeit securities.

These use cases and success stories demonstrate blockchain technology's promise for combating fraud in financial markets. These solutions seek to improve security, trust, and efficiency in a variety of financial operations by leveraging blockchain capabilities such as transparency, immutability, and decentralized consensus.

7.6 CONCLUSION AND FUTURE PROSPECTS OF BLOCKCHAIN TECHNOLOGY FOR FRAUD PREVENTION IN THE FINANCIAL MARKET

Blockchain technology has a potential future for preventing fraud in the financial system. Blockchain's intrinsic qualities, such as immutability, cryptographic hashing, and decentralized consensus, form a solid foundation for improving financial sector security. Using blockchain, financial organizations may generate tamper-proof and transparent transaction records, lowering the risk of fraud, illegal access, and data tampering. Blockchain-based identity management solutions have the potential to transform how identities are validated and managed. Self-sovereign identity systems on the blockchain provide individuals control over their personal information, lowering the danger of identity theft and fraud. Financial institutions can use these technologies to improve KYC processes, simplify customer onboarding, and prevent identity fraud. Smart contracts, or self-executing agreements with predetermined conditions inscribed on the blockchain, can automate and enforce corporate operations in a transparent and tamper-proof manner. This automation can improve financial operations while lowering the risk of fraud caused by human error or malevolent intent. Smart contracts allow for automated transaction verification, validation, and execution, which improves efficiency and reduces fraud. Blockchain's transparency and immutability allow for real-time audits and compliance monitoring. Financial transactions recorded on the blockchain are easily audited, resulting in a transparent and verifiable trail. This can aid in the detection of fraudulent activity, ensuring regulatory compliance, and reducing the need for time-consuming and laborious auditing procedures.

Blockchain's decentralized and distributed nature makes it extremely resistant to data tampering and manipulation. By storing financial data on blockchain, financial organizations can assure the integrity and immutability of crucial information. This lowers the likelihood of fraudulent activity, such as faking transaction records, changing contract conditions, or manipulating financial statements. Blockchain enables secure and transparent communication amongst various parties in the financial sector. Financial institutions can share and verify data on a single blockchain network, allowing for more efficient and coordinated fraud prevention measures. Shared databases, paired with privacy-preserving techniques such as zero-knowledge proofs, can aid in identifying patterns of fraudulent conduct while maintaining data privacy.

While blockchain technology has enormous potential for fraud prevention in the financial system, there are several issues that must be solved. These include scalability, legal frameworks, interoperability, and blockchain interaction with current financial systems. Overcoming these issues will be critical for widespread adoption and realizing blockchain's potential in preventing financial fraud. Overall, blockchain technology has the potential to revolutionize fraud prevention in the financial sector by improving security, transparency, and trust. Continued research, development, and collaboration among stakeholders will define blockchain technology's future potential for preventing financial fraud.

KEYWORDS

- **blockchain**
- **decentralization**
- **distributed ledger technology**
- **financial market**
- **financial system**
- **fraudulent**

REFERENCES

Amarasinghe, T., Aponso, A., & Krishnarajah, N. (2018). Critical analysis of machine learning based approaches for fraud detection in financial transactions. *Proceedings of the International Conference on Machine Learning Technologies (ICMLT)*, 12–17.

Assefa, S. A., Dervovic, D., Mahfouz, M., Tillman, R. E., Reddy, P., & Veloso, M. (2020). Generating synthetic data in finance: Opportunities, challenges and pitfalls. *Proceedings of the 1st ACM International Conference on AI in Finance*, (44), 1–8.

Aziz, A., Goldman, H. M., & Olsen, N. (2007). Facets of Type A personality and pay increase among the employees of fast-food restaurants. *International Journal of Hospitality Management, 26*(3), 754–758.

Beena, F., Mearaj, I., Shukla, V. K., & Anwar, S. (2021, February). Mitigating financial fraud using data science— 'A case study on credit card frauds.' In *Proceedings of the International Conference on Innovative Practices in Technology and Management (ICIPTM)*, Noida, India.

Bhaskar, P., Tiwari, C. K., & Joshi, A. (2020). Blockchain in education management: Present and future applications. *Interactive Technology and Smart Education, 18*(1), 1–17.

Chen, Y., & Bellavitis, C. (2020). Blockchain disruption and decentralized finance: The rise of decentralized business models. *Journal of Business Venturing Insights, 13*.

Christidis, K., & Devetsikiotis, M. (2016). Blockchains and smart contracts for the Internet of things. *IEEE Access, 4*, 2292–2303.

Cruz, J. P., Kaji, Y., & Yanai, N. (2018). RBAC-SC: Role-based access control using smart contract. *IEEE Access, 6*, 12240–12251.

Engelhardt, M. A. (2017). Hitching healthcare to the chain: An introduction to blockchain technology in the healthcare sector. *Technology Innovation Management Review, 7*(10), 22–34.

Goldsmith, R. W. (1969). *Financial Structure and Development*. Yale University Press.

Gyimah-Brempong, K. (2002). Corruption, economic growth, and income inequality in Africa. *Economics of Governance, 3*(3), 183–209.

Haque, A. B., Islam, A. K. M. N., Hyrynsalmi, S., Naqvi, B., & Smolander, K. (2021). GDPR compliant blockchains—A systematic literature review. *IEEE Access, 9*, 50593–50606.

Hardouin, P. (2009). Banks governance and public-private partnership in preventing and confronting organized crime, corruption and terrorism financing. *Journal of Financial Crime*.

Hofmann, F., Wurster, S., Ron, E., & Bohmecke-Schwafert, M. (2017). The immutability concept of blockchains and benefits of early standardization. In *Proceedings of the ITU Kaleidoscope: Challenges for a Data-Driven Society (ITU K)*, Nanjing, China, 1–8.

Hwa, C. S. (2016). Legal issues for the introduction of distributed ledger based on blockchain technology focused on the financial industry. *Korea Financial Law Association, 13*(2), 107–138.

Hyvarinen, H., Risius, M., & Friis, G. (2017). A blockchain-based approach towards overcoming financial fraud in public sector services. *Business & Information Systems Engineering, 59*(6), 441–456.

Iqbal, M., & Matulevičius, R. (2021). Exploring sybil and double-spending risks in blockchain systems. *IEEE Access, 9*, 76153–76177.

Keefer, P., & Knack, S. (1997). Why don't poor countries catch up? A cross-national test of an institutional explanation. *Economic Inquiry, 35*, 590–602.

Khelil, I., Khlif, H., & Amara, I. (2021). Political connections, political corruption and auditing: A literature review. *Journal of Financial Crime*.

Kim, H. M., & Laskowski, M. (2018). Toward an ontology-driven blockchain design for supply-chain provenance. *Intelligent Systems in Accounting, Finance and Management, 25*(1), 18–27.

Kiviat, T. I. (2015). Beyond Bitcoin: Issues in regulating blockchain transactions. *Duke Law Journal, 65*(3), 569–608.

Kraft, D. (2016). Difficulty control for blockchain-based consensus systems. *Peer-to-Peer Networking and Applications, 9*(2), 397–413.

Lee, S. H., & Kim, K. (2016). Device authentication in smart grid system using blockchain. KAIST, Daejeon.

Levine, R. (2003). More on finance and growth: More finance and growth: More finance, more growth? *Review, Federal Reserve Bank of St. Louis*, 31–96.

Li, H., Xu, L., & Zou, H. (2000). Corruption, income distribution and growth. *Economics and Politics, 12*, 155–182.

Lim, W. M. (2022). Ushering a new era of global business and organizational excellence: Taking a leaf out of recent trends in the new normal. *Global Business and Organizational Excellence, 41*(5), 5–13.

Lim, W. M., Srivastava, S., Jain, A. K., Malik, N., & Gupta, S. (2023). When employees feel betrayed: The mediating role of psychological contract violation on nepotism and workplace commitment in the hotel industry. *International Journal of Hospitality Management, 108*, 103381.

Lu, J., Li, W., Wang, Q., & Zhang, Y. (2020). Research on data quality control of crowdsourcing annotation: A survey. In *Proceedings of the IEEE International Conference on Dependable, Autonomic and Secure Computing; International Conference on Pervasive Intelligence and Computing; International Conference on Cloud and Big Data Computing; International Conference on Cyber Science and Technology Congress (DASC/PiCom/CBDCom/CyberSciTech)*, Calgary, AB, Canada, 201–208.

Lucey, B. M. (2023). Corruption in finance research: The state of art and future research agenda. *Journal of Economic Criminology, 1*, 100001. https://doi.org/10.1016/j.jeconc.2023.100001 (accessed on 26th July 2025).

McGhin, T., Choo, K. K. R., Liu, C. Z., & He, D. (2019). Blockchain in healthcare applications: Research challenges and opportunities. *Journal of Network and Computer Applications, 135*, 62–75.

Méon, P., & Sekkat, K. (2005). Does corruption grease or sand the wheels of growth? *Public Choice, 122*, 69–97.

Mo, P. (2001). Corruption and economic growth. *Journal of Comparative Economics, 29*, 66–97.

Nakamoto, S. (2008). *Bitcoin: A Peer-To-Peer Electronic Cash System*. https://bitcoin.org/bitcoin.pdf (accessed on 26th July 2025).

Niranjanamurthy, M., Nithya, B. N., & Jagannatha, S. (2019). Analysis of blockchain technology: Pros, cons and SWOT. *Cluster Computing, 22*(S6), 14743–14757.

Nofer, M., Gomber, P., Hinz, O., & Schiereck, D. (2017). Blockchain. *Business & Information Systems Engineering, 59*(3), 183–187.

O'Dair, M., & Beaven, Z. (2017). The networked record industry: How blockchain technology could transform the record industry. *Strategic Change: Briefings in Entrepreneurial Finance, 26*(5), 471–480.

Online Payment Fraud Losses to Exceed $206 Billion Over the Next Five Years; Driven by Identity Fraud. (2022). *Juniper Research*. https://www.juniperresearch.com/press/online-payment-fraud-losses-exceed-206-bn (accessed on 26th July 2025).

Paech, P. (2017). The governance of blockchain financial networks. *Modern Law Review, 80*(6), 1073–1110.

Pranto, T. H., Hasib, K. T. A. M., Rahman, T., Haque, A. B., Islam, A. K. M. N., & Rahman, R. M. (2022). Blockchain and machine learning for fraud detection: A privacy-preserving and adaptive incentive-based approach. *IEEE Access, 10*, 87115–87134.

Radanovic, I., & Likic, R. (2018). Opportunities for use of blockchain technology in medicine. *Applied Health Economics and Health Policy, 16*(5), 583–590.

Rajan, R. G., & Zingales, L. (1998). Financial dependence and growth. *American Economic Review, 88*(3), 559–586.

Rikap, C., & Lundvall, B. (2021). Tech giants and artificial intelligence as a technological innovation system. In *The Digital Innovation Race* (pp. 65–90). Springer. https://doi.org/10.1007/978-3-030-89443-6_4 (accessed on 26th July 2025).

Rizal Batubara, F., Ubacht, J., & Janssen, M. (2019). Unraveling transparency and accountability in blockchain. *Proceedings of the 20th Annual International Conference on Digital Government Research*, 204–213.

Roriz, R., & Pereira, J. L. (2019). Avoiding insurance fraud: A blockchain-based solution for the vehicle sector. *Procedia Computer Science, 164*, 211–218.

Savelyev, A. (2018). Copyright in the blockchain era: Promises and challenges. *Computer Law & Security Review, 34*(3), 550–561.

Schnatterly, K., Gangloff, K. A., & Tuschke, A. (2018). CEO wrongdoing: A review of pressure, opportunity, and rationalization. *Journal of Management, 44*(6), 2405–2432.

Sedgo, H., & Omgba, L. D. (2022). Corruption and distortion of public expenditures: Evidence from Africa. *International Tax and Public Finance*, 1–34.

Shahbazi, Z., & Byun, Y. C. (2021). Integration of blockchain, IoT and machine learning for multistage quality control and enhancing security in smart manufacturing. *Sensors, 21*(4), 1467.

Soonduck, Y. (2017). Blockchain-based financial case analysis and its implications. *Asia Pacific Journal of Innovation and Entrepreneurship, 11*(3), 312–321.

Szabo, N. (1997). Formalizing and securing relationships on public networks. *First Monday, 2*(9). https://firstmonday.org/article/view/548/469 (accessed on 26th July 2025).

Vargas-Hernández, J. G. (2009). The multiple faces of corruption: Typology, forms and levels. In *Organizational Immunity to Corruption: Building Theoretical and Research Foundations* (pp. 111–117).

Varma, J. R. (2019). Blockchain in finance. *Vikalpa, 44*(1), 1–11.

Venard, B., & Hanafi, M. (2008). Organizational isomorphism and corruption in financial institutions: Empirical research in emerging countries. *Journal of Business Ethics, 81*(2), 481–498.

Xu, M., Chen, X., & Kou, G. (2019). A systematic review of blockchain. *Financial Innovation, 5*, 27. https://doi.org/10.1186/s40854-019-0147-z (accessed on 26th July 2025).

Yang, C. (2022). Incremental outlier feature clustering algorithm in blockchain networks based on big data analysis. *IETE Journal of Research*, 1–9.

Zheng, W., Zheng, Z., Chen, X., Dai, X., Li, P., & Chen, R. (2019). NutBaaS: A blockchain-as-a-service platform. *IEEE Access, 7*, 134422–134433.

CHAPTER 8

Impact of Blockchain Technology in the Financial Market

SUBHADRA RAJPOOT*

Amity University, Greater Noida, Uttar Pradesh, India

*Corresponding author

ABSTRACT

Blockchain is a distributed ledger system that uses a decentralized consensus protocol to record transactions securely over a network of computers. Unlike established centralized systems, blockchain runs on a peer to-peer network with every participant (node) having access to the entire database and its full history. This, then, has the advantage of making the system more resilient to corruption or hacking. The ledger (transaction history) is replicated across multiple participants on the hardware framework. Transparency: All participants have the same copy of the blockchain (without a need for a central authority). Autonomous agreements are contracts that execute themselves because their content is directly encoded in them. They apply and effectuate the conditions of a contract without intermediaries when certain predefined stipulations come into play. The decentralized and secure method in which cryptocurrencies operate is through the mechanism of blockchain, which is widely acknowledged. It can simplify the process, decrease fraud, and make it safer. This study pursues a comprehensive survey of the diversified use cases of blockchain throughout the global financial ecosystem. This chapter is an effort to shed light on the Strengths, Weaknesses, Opportunities, and Threats of blockchain technology in financial services.

FinTech and Blockchain Technology in Financial Markets. Neha Puri, Anjali Munde, Emilia Alaverdov (Eds.)
© 2027 Apple Academic Press, Inc. Co-published with CRC Press (Taylor & Francis)

DOI: 10.1201/9781779641298-8

8.1 INTRODUCTION

Important parts of the financial industry are currently undergoing major changes. The long-term continuous development of cutting-edge digital technologies has significantly contributed to the increase in the number of service options currently available. As digitalization makes traditional business practices obsolete, it has made it possible for new market entrants to get in (Feyen et al., 2021). OECD recently released a report examining the impact of several new and innovative data-oriented technologies on various operations and services around the world in the financial sector (e.g., payment transactions, financial investments, loans, etc.). Basics of virtual currencies such as Bitcoin. Cryptocurrency and its underlying blockchain technology are being touted as the next big thing after the birth of the Internet. One area where these technologies are expected to have a major impact is in the financial sector. The distributed ledger technology is an example of the blockchain, which can change how traditional financial institutions operate by cutting their costs, speeding up transaction processing times, increasing transparency, and providing information for auditing purposes, among other merits. There is a gender. Cryptocurrency promises a new native digital asset class with no central authority (Gompert al., 2021; Ilk et al., 2021).

Blockchain relies on transactions taking place in traditional corporate databases, which underpin nearly all financial services functions. It has the potential to replace any business activity: financial transactions with low transparency and limited traceability are vulnerable to disruption by blockchain applications.

Although the primary purpose of blockchain technology was to make cryptocurrency generation easier, it now has many more applications than that. Blockchain, or distributed ledger technology, opens the door for the safe and middleman-free transmission of digital assets, money, and information (Swan, 2019). Blockchain is becoming a new area of study in the financial industry and academia, with applications ranging from supply chain finance, international payments, and cryptocurrencies (Goldstein et al., 2019). According to Davidson et al., the decentralized structure of blockchain technology is important because there is no intermediary or counterparty risk and has a significant impact on institutional economics (2016). Application of blockchain technology is allowed in the development of operational tools and new services in the future. Banks and other financial institutions are striving to integrate cutting-edge fintech technologies, such as blockchain, to remain competitive in the increasingly competitive global financial sector. Volchenkov et al. (2017) show that an increasing global count of commercial banks is adopting blockchain technology to reduce the cost of interaction between economic entities, provide data transparency, and manage and minimize financial transaction costs. The use of digital technology like blockchain will pay off in the long run. According to Vega and Clohessy (2021), the implementation of blockchain technology can enhance the security of financial transactions involving parties with dubious reputations by doing away with the necessity for third parties to act as mediators.

However, some parts of the financial sector believe that blockchain technology will have a small impact on our daily lives. The concept of Bitcoin was created by financial institutions as an alternative. This clearly contradicts the fact that most Bitcoin transactions currently take place on exchanges, where cryptocurrencies and bitcoin can be exchanged for traditional fiat currencies. Various risks identified by financial markets hinder widespread acceptance in global financial markets. Blockchain technology is the main foundation of cryptocurrencies like Bitcoin, Ethereum, and others, which have drawn harsh criticism from conventional financial institutions. Many nations in Asia, Europe, and other parts of the world have shown their unwillingness to support the development and adoption of financial legislation for the financial sector's use of blockchain technology due to the associated dangers.

This chapter offers a theoretical viewpoint by using a SWOT analysis to comprehend the possible advantages and difficulties that could affect the adoption of blockchain technology. Potential advantages and challenges of blockchain applications are covered in "Potential Advantages" and "Challenges of Blockchain Applications in the Financial Sector," respectively.

8.2 BACKGROUND – BLOCKCHAIN TECHNOLOGY

When using typical database technology to record financial transactions, there are a number of challenges. Consider the selling of a house as an illustration. After money is exchanged, the buyer becomes the owner of the item. The financial transactions can be separately recorded by the buyer and the seller, but neither source is trustworthy. Both the buyer and the seller might easily argue that they have paid the money even though they haven't received it, or that they haven't received it at all. A trustworthy third party must oversee and approve transactions to avoid potential legal issues. The presence of this central authority introduces another level of risk to the transaction in addition to making it more complicated.

The popularity of blockchain has grown because of Bitcoin's ascent. But this technology is not just used in the banking sector. Blocks of cryptocurrency connected by chains were the initial definition of a blockchain. FinTech has given this novel idea a lot of attention (Mu, 2016). To add blocks to the chain without the assistance of a reliable third party, Satoshi Nakamoto invented the first Blockchain in 2008 and employed a technique akin to hash cash (Narayanan et al., 2016). The way consumers interact with businesses is being revolutionized by blockchain, a rapidly developing financial technology. Blockchain establishes an impermeable, decentralized transaction record system, which lessens these issues. Blockchain creates a single ledger for the vendor and the buyer in the context of real estate transactions. Every transaction needs the consent of both parties and is automatically updated in real time in the ledgers of both parties.

Although blockchain has been seen as a new foundation for global transactions, it receives attention as the technology underlying Bitcoin and other cryptocurrencies (Nguyen, 2016; Staples et al., 2017). According to Yoo (2017), a blockchain is an unalterable, distributed, and continuous account database. Blockchain's decentralized system backed by an extensive security chain is its best feature. The distributed trust that comes from Blockchain technology is the key innovation; it minimizes the need for a reliable third party to mediate transactions, lowers trading costs, and shortens transaction times (Staples et al., 2017).

Blockchain allows numerous nodes to maintain data integration, as opposed to a single node transaction cash book. A copy of the most recent transaction is sent to every node on the blockchain. The recently completed transactions are combined into a single block, and the node that gets to transmit its block is chosen by the so-called consensus process (which is covered in more depth below). Every transaction in this block is evaluated by other nodes to see whether it is valid or not. This block and the one after it will refer to approved blocks. Figure 8.1 that follows explains the three block structure blockchain mechanism.

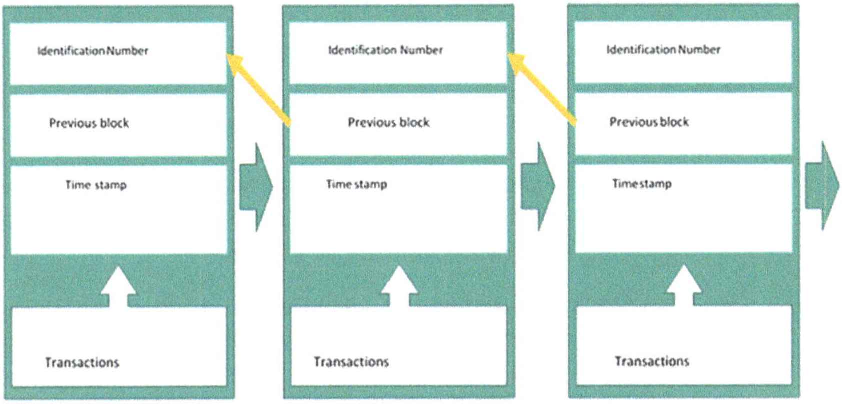

FIGURE 8.1 Three block structure blockchain mechanism's.

They are constructed using asymmetric cryptography so that their legitimacy can be confirmed. Anybody on the network can verify that the transaction originated from the sender by decrypting it using the sender's public key. The primary objective is to guarantee the transaction's validity and integrity, preventing transaction manipulation and enabling the sender to be identified at any moment. It was demonstrated how a significant amount of consensus knowledge is used to manage which node will transmit its block to the. The so-called miners, or nodes that collect new transactions, must first examine and encode the new transactions before they can add the next block to the chain. Payment is then made to ensure that enough nodes take part in this phase. Nodes should only construct blocks that contain real transactions that don't conflict with each other or the rest of the blockchain. No transaction that uses the same asset more than once should be included, according to Yaga et al. (2018). The most extensively utilized consensus method, proof-of-work, is what the Bitcoin network is dependent on. Sobti and Geetha (2012) state that a hash function is used by the 256-bit Bitcoin blockchain to compress a significant amount of data into a fixed-size number called the hash value. On the Bitcoin blockchain, a maximum value is now defined by the number of leading zeros in a hash value. This is accomplished by appending a time element to a data point. Currently, the hash value is being adjusted through a trial-and-error procedure until the requirements are satisfied. The number of attempts to find a workable solution per unit of time is measured by the hash rate (Rutland, 2018). Lastly, every other node can verify that the task has been finished. To verify that the hash value satisfies the requirements, all that has to be done is input the block containing the nonce into the hash function (Zheng et al., 2017).

8.3 IMPORTANT FEATURES OF BLOCKCHAIN

The four primary features of blockchain are as follows:

1. **Decentralization:** According to Zheng et al. (2018), every transaction in a conventional centralized transaction system must be approved by a central trusted entity, such as the central bank. Every party on the blockchain has access to the database and may see the transaction history (Tapscott and Tapscott, 2017). This chain's primary benefit is its widespread network replication. Therefore, for a criminal or abusive government agency to avoid detection, all copies of the blockchain must be changed at the same time. Additionally, distributed ledgers minimize the possibility of fraud by automatically and instantly recording transactions (Rennock et al., 2018). Decentralized infrastructures have the potential to be highly successful in handling blockchain-related activities due to their limited border conditions.
2. **Anonymity of Users:** Blockchain addresses are used for transactions. According to Tapscott and Tapscott (2017), every user on a blockchain has an individual alphanumeric address that they can choose to keep private or public. According to Zheng et al. (2018), users can engage with the blockchain network using the created address, and their private information is no longer stored by a third party. Certain privacy is preserved by this method, but blockchain cannot provide absolute privacy protection because of built-in limitations.
3. **Consensus Mechanism:** A consensus mechanism is incorporated into the network as there isn't a single trusted agent throughout it. Its goal is to come to a consensus on the validity of each and every record. Gaining control over more than 51% of the network's accounting nodes can be used to fabricate a record that doesn't exist. Therefore, it is simple to identify any distortion (Huang et al., 2019).

4. **Execution:** According to Tapscott and Tapscott (2017), users can use rules and algorithms to start transactions between nodes. If specific requirements are satisfied, blockchain can also carry out programming. We refer to this as a smart contract. "It is not just one or two characteristics of blockchain that render this technology creative and attract individuals' attention," states H, the managing director of a blockchain company. This new technology is valuable because it incorporates blockchain properties like immutability, anonymity, and decentralization.

8.4 BLOCKCHAIN'S ADVANTAGES AND PROSPECTS FOR THE WORLD FINANCIAL SECTOR

Blocks 1.0, 2.0, and 3.0 are the three technological phases that blockchain technology has undergone. Bitcoin is a successful project of Block 1.0, which is a currency. Block 2.0 encompasses smart contracts, bonds, mortgages, loans, and futures in addition to cash transactions. Government, research, health, culture, the arts, and literacy all leverage blockchain 3.0 as a platform and application (Swan, 2015). Examining the potential applications of Blockchain technology across the financial and banking sectors, Distribution Ledger Technologies—along with Bitcoin and other cryptocurrencies—offer financial institutions a wide range of operational benefits, such as enhanced security, cost savings, and immutability. Additionally, because database incidents are less common on blockchains than they are on databases, it has been found that putting data on a blockchain provides higher security than storing it in a single database. As a result, blockchain technology helps customers with data transparency when it is used in an industry that needs information releases. Blockchain technology thus has the potential to be used in the financial services industry. Its variety of applications is projected to expand soon. Cost-cutting measures are always important to consider when implementing blockchain technology. Research has shown that blockchain technology has the potential to significantly reduce infrastructure costs, which can save money over the course of a fiscal year.

Blockchain is a revolutionary technology that encourages data storage transparency. Because it is an unchangeable ledger, trust in the information it contains is increased. According to Queiroz and Wamba (2019), one of the main factors influencing people's decision to embrace blockchain technology is the high level of transparency it offers its users. They went on to say that Blockchain enables all network participants to transact in a way that fosters confidence while keeping them informed of pertinent facts. A 2019 study by Clohessy and Acton found that reducing expenses, enhancing data security, improving operational and transactional efficiency, and promoting transparency in transactional processes are the main drivers of blockchain technology adoption. In 2020, Reinhardt, Oliveira, and Ring believe the capabilities of blockchain technology will improve financial institutions' operational quality and efficiency even more. Using Blockchain technology, Berentsen and Schar (2018) illustrated the transparency of cryptocurrency transactions (Figure 8.2).

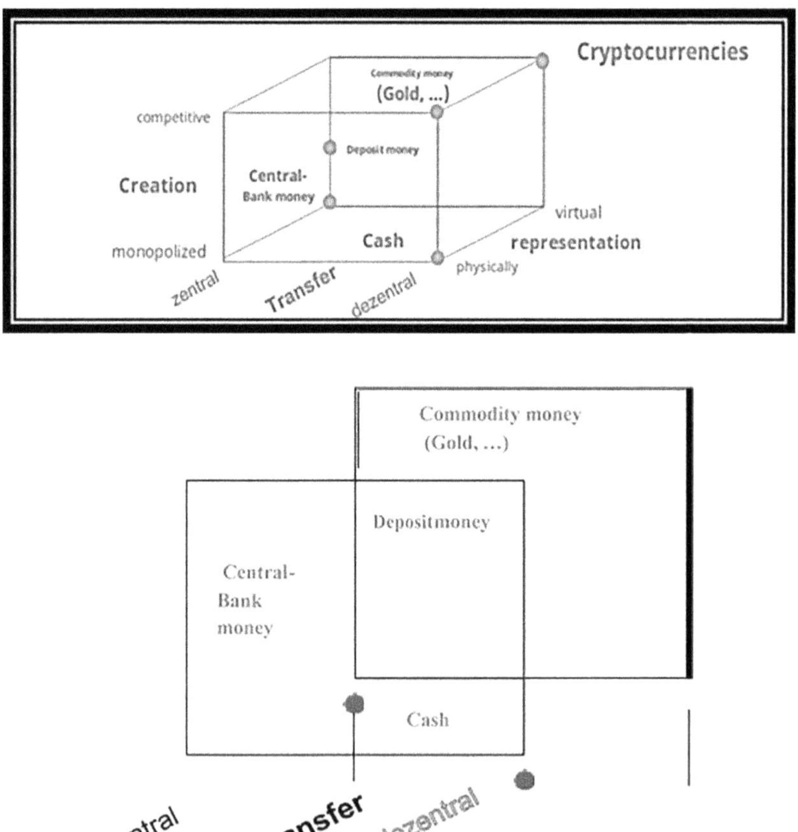

FIGURE 8.2 Dimensions of payment systems.

Berentsen and Schar (2018) used three dimensions to describe currencies or general payment systems. According to Kevin and Suresha (2018), these are the regulations that control the formation, representation, and transfer of monetary value. For instance, the creation of money can be monopolized by a central bank or done in a competitive way, as in the case of gold mining. Having money physically represented as cash or deposited into an account are two ways to symbolize money. When describing currencies or general payment systems, it is typical to consider several factors that contribute to a thorough understanding. Here are three crucial factors to think about are discussed in the subsections.

8.4.1 FUNCTIONALITY AND USE CASES

1. **Exchange Medium:** Examine the way the currency or payment system promotes exchanges. Evaluate how easily it can be used for day-to-day transactions and international trade.
2. **Value Store:** Examine the ability of the currency to retain its value over time. Stable currencies are generally considered good stores of value, while those with high volatility may not be suitable for long-term savings.
3. **Account Unit:** Evaluate whether the currency is a widely accepted unit of measurement for goods, services, and assets. A stable unit of accounts simplifies pricing and economic planning.

8.4.2 INFRASTRUCTURE AND TECHNOLOGY

1. **Blockchain Along with Distributed Ledger Concepts:** Consider whether the currency or payment system is based on blockchain or distributed ledger technology. This affects aspects like transparency, security, and decentralization.
2. **Transaction Speed and Scalability:** Evaluate how quickly transactions can be processed and whether the system can scale to handle a growing number of transactions. This is crucial for mainstream adoption and efficient payment processing.

8.4.3 REGULATORY ENVIRONMENT AND GOVERNANCE

1. **Legal Status:** Consider the legal standing of the currency or payment system in various jurisdictions. Some currencies may be widely accepted and regulated, while others operate in a legal gray area or are explicitly banned.
2. **Regulatory Compliance:** Examine whether the currency or payment system complies with existing financial regulations. This includes anti-money laundering (AML) and knowing your customer (KYC) regulations.
3. **Governance Structure:** Evaluate the governance model of the currency or payment system. Is it decentralized, governed by a community, or controlled by a central authority? The governance structure can impact decision-making, updates, and overall system stability.

These dimensions provide a broad framework for assessing currencies or payment systems, but it's important to note that the cryptocurrency and payment landscape is dynamic, and new factors may become relevant over time.

People can exchange cash in two different ways: directly through a bank or, in the case of a cash exchange, through another method. According to Reyna et al. (2018), cryptocurrencies like Bitcoin and Ethereum combine decentralized transfer and virtual representation, which are beneficial from an economic perspective and provide notable efficiency advantages over the physical one. The combination of virtual and decentralized transmission was not possible prior to the development of Blockchain.

Furthermore, Blockchain can help prevent excessive price volatility by allocating the money supply to a small group of individuals. The large swings in the value of cryptocurrencies prompted the creation of stable coins. The fundamental idea is that there is a fixed amount of money behind the currency. Stable coins, by virtue of their design, are dependent on a central authority, namely the organization or group that provides the collateral (Diaby et al., 2021). That directly contradicts the original intent of cryptocurrencies, which was to do away with the requirement for centralized authority.

The creation of Smart Contracts is a key use of blockchain technology in the finance sector. Chenthara et al. (2020) tracked the evolution of.

The phrase "smart contract" refers to a section of code that is kept on the blockchain and that, upon fulfillment of specific requirements, is automatically carried out without the original author's intervention. The blockchain can store more data and do calculations thanks to smart contracts. Any stakeholder at any time can verify any information contained in a smart contract. Because of blockchain technology, once a code is recorded, it cannot be altered. The code's ability to be read by machines guarantees that its meaning cannot be unclear. Because the terms of the contract cannot be changed afterward and are sometimes encrypted using cryptography, they offer a high level of security. In most cases, a blockchain performs instantaneous payments or item transfers. Extreme event insurance is one application. After a thorough investigation, Zheng et al. (2020) assert that using Blockchain instead of traditional techniques can help expedite the claims procedure. The distribution of a crypto asset or the execution of complex payment operations have both used the Bitcoin blockchain (Figure 8.3).

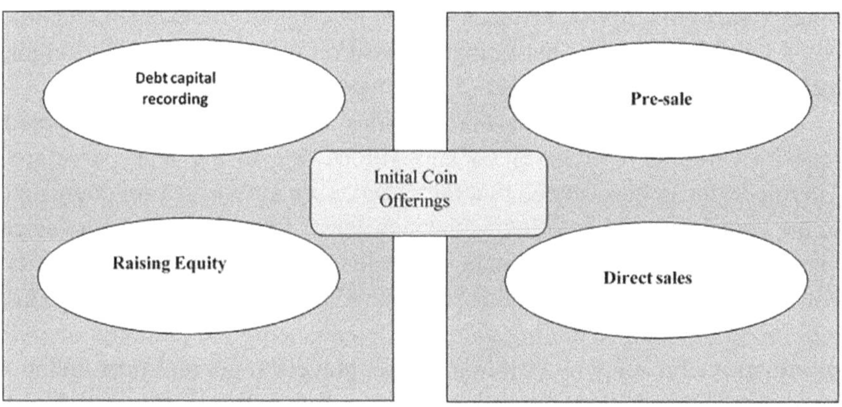

FIGURE 8.3 Types of initial coin.

Blockchain technology, which creates consensus on completed transactions, can be used to record non-tangible assets. In this sense, it is possible to do equity-related transactions on a blockchain. The blockchain requirements for this application are different from those for cryptocurrencies since securities transactions are subject to tax, regulatory, and privacy restrictions. Tracing a security's ownership would be easy. Smart contracts can also be used to automate processes like dividend and interest payments.

This new system would have far more efficient processing compared to the one that now exists. A considerably faster settlement, or perhaps real-time, might be used in place of the current two trading days after the trading day. If payment and security delivery were completed concurrently, settlement concerns might be minimized or completely removed. There would be no need for a central counterparty (Schar, 2021). Based on the proof-of-work methodology, Chiu and Koeppl (2019) have constructed a theoretical model of a public, permissioned blockchain (Yermack, 2017). Market participants have the option to pay a fee, which makes their transaction more appealing to miners and expedites the completion of the transaction. More flexibility in transaction processing speed is made possible by this. Therefore, to provide incentives to the miners, the money received from the fees is needed (Huibers, 2021). Smart contracts built on blockchain technology enable fully automated, real-time financial derivative transactions. In addition to lowering counterparty risk, it has the drawback of requiring market participants to keep liquidity on hand if a margin call is likely at any time (Yaga et al., 2018). Yermack (2017) investigates blockchain technology's ramifications. His reasoning is based on the notion that blockchain technology supports every facet of securities trading.

As mentioned earlier in the previous section, blockchain-enabled securities trading will never be anonymous due to the legal structure. Therefore, it can be assumed that the supervisory authorities are cognizant of the shareholders' identities, at the very least. This would make it easier to spot insider trading and make it harder to acquire holdings covertly without having to comply with regulatory reporting requirements. This will minimize insider trading gains while enhancing the market's overall information efficiency by enabling market players to get real-time updates on reportable securities transactions by corporate insiders (directors deals) (Sharma and Carpathia, 2015). If stock options were likewise built on a blockchain, re-dating them for management would be more challenging (Singh et al., 2020).

8.5 BLOCKCHAIN CHALLENGES IN FINANCIAL MARKETS

Blockchain technology has become a disruptive force in the banking industry, promising more security, efficiency, and transparency. However, financial institutions face several dangers and problems in addition to its potential benefits. This essay examines the main dangers and difficulties related to blockchain adoption in the financial industry and provides advice on how to get beyond them (Chen and Wang, 2018; Lin and Liao, 2017). Banks using blockchain technology face several risks, including reputational, legal, regulatory, and operational concerns (Bauer and Hein, 2006; Deshpande et al., 2017). Blockchain technology, originally devised for the digital currency Bitcoin, has rapidly gained traction across various industries, particularly in finance. Its decentralized and immutable nature has the potential to revolutionize traditional financial processes, from payments and settlements to regulatory compliance and identity verification. However, the implementation of blockchain in the financial sector is not without its hurdles Pennathur (2001). Few of the challenges of blockchain technology are:

1. **Scalability:** Because of their low transaction throughput, blockchain networks—particularly open ones like Bitcoin and Ethereum—have scalability problems. As financial institutions scale their operations, they need blockchain solutions capable of handling high transaction volumes without compromising speed or cost-effectiveness (Deshpande et al., 2017).
2. **Interoperability:** The lack of interoperability between different blockchain platforms and legacy systems complicates the integration of blockchain into existing financial infrastructure. Financial institutions may struggle to ensure seamless data exchange and communication between disparate blockchain networks and traditional databases.
3. **Security:** Smart contract vulnerabilities, private key management issues, and the risk of 51% attacks are among the security challenges that financial institutions must address when deploying blockchain solutions (Mohanta et al., 2019).
4. **Privacy and Confidentiality:** Balancing the transparency of blockchain with the privacy and confidentiality requirements of financial transactions is a delicate task. Financial institutions must implement privacy-enhancing techniques, such as zero-knowledge proofs and secure multiparty computation, to protect sensitive information while leveraging blockchain technology.

5. **Risks: Regulatory Compliance:** Non-compliance with evolving regulatory standards can result in hefty fines, reputational damage, and legal repercussions for financial institutions leveraging blockchain. Regulatory uncertainty surrounding cryptocurrencies and initial coin offerings (ICOs) poses a risk of regulatory crackdowns and enforcement actions (Salimitari et al., 020). The integration of blockchain into existing financial systems may cause operational disruptions, including downtime, data inconsistencies, and transaction delays. Financial institutions must carefully plan and execute blockchain implementations to minimize the risk of operational disruptions and ensure business continuity. Public perception of blockchain technology, particularly in the context of its association with cryptocurrencies and illicit activities, can impact the reputation of financial institutions. Instances of security breaches, fraudulent transactions, or regulatory violations involving blockchain-based systems can tarnish the reputation of financial institutions and erode customer trust. The rapid pace of technological innovation in the blockchain space introduces the risk of investing in outdated or incompatible solutions Harwood-jones (2016) and Khan and Salah (2018). Financial institutions must stay abreast of emerging blockchain technologies and standards to mitigate the risk of technological obsolescence and ensure long-term viability. While blockchain holds immense potential to transform the financial sector, its adoption is accompanied by various challenges and risks that demand careful consideration and strategic planning.

8.6 BLOCKCHAIN ADOPTION, APPLICATIONS, AND FUTURE OUTLOOK IN FINANCIAL MARKETS

Blockchain technology started with Bitcoin in 2008, then several other cryptocurrencies, and in 2015, open source Hyperledger added even more functionality to the system. This movement has the ability to upend the global finance industry. Blockchain technology is constantly evolving, and these developments go far beyond the cryptocurrency speculation of investors. Blockchain presents both many potential and challenges. These days, there are blockchain applications that may be used to execute financial transactions and clear exchanges of a variety of financial assets.

For instance, HSBC recently declared the successful completion of a blockchain-powered trade finance transaction. In a matter of minutes, the transaction was finished. Contrast this with a comparable transaction that may take up to a week to process through a paper-based system. International trade financing will look like this in the future.

These are a few significant blockchain obstacles:

1. **Performance Issues:** With Bitcoin, transaction speeds are extremely slow—three to seven transactions can be completed in a second. Though it was designed to speed up transactions, the Ethereum cryptocurrency blockchain can only process 15 transactions per second. In contrast, consider the fact that traditional payment processing systems handle tens of thousands of transactions every second.
2. **Cybercrime and Cryptocurrency Hacking:** Bitcoins have a strong link to illegal activity. Additionally, the value of bitcoin exchanges was lost by illegal theft via hacking, amounting to around $1 billion. For these and other reasons, Hyperledger applications are separating from cryptocurrencies. However, regarding security, there are still a lot of unsolved concerns and dislike for the connection to crime. Hackers cannot affect decentralized, peer-to-peer public encrypted blockchains; on the other hand, private blockchains that have centralized management, such as bitcoin exchanges, are very susceptible to hackers.
3. **Standards:** There is an issue with blockchain system interoperability. Over several decades, legacy systems evolved through increased interconnectivity. To gain more traction, blockchain technology must likewise adhere to this broad interoperability pattern.
4. **Transaction Costs:** A significant issue with several blockchain technologies is the expense of the electricity and computer servers required to handle transactions. Bitcoins are produced by "mining." In many places today, mining bitcoins is a difficult computational challenge that consumes more electricity than it is worth. The cost of a transaction on Ethereum is roughly 15 cents.
5. **Price Volatility in Cryptocurrency:** Cryptocurrencies are less preferred for long-term wealth storage due to their volatile price

swings, which resulted in an abrupt 80% loss of value in 2018. This extreme value volatility undermines any system that accepts cryptocurrency payments.

6. **Regulations:** Governments everywhere are attempting to keep up with the times by regulating this technology. In many instances, such as China and South Korea, they are even outlawing cryptocurrency and prohibiting exchanges. The largest blockchain issues are coming from the growing regulation of this technology. This will impede the implementation of new worldwide blockchain/ Hyperledger systems.
7. **Acceptance:** For blockchain technology to become more widely accepted, banking and industry sectors must collaborate. The fact that this threatens monopolies, like those in central bank-dominated currency exchange markets, is one of the main obstacles.
8. **Conversion from Legacy Systems:** A major obstacle to implementing blockchain technology in place of outdated systems is the widespread aversion to change. This expense could entail more than just a monetary outlay. In certain industries, such as international trade finance, Hyperledger blockchain technology is only now being utilized for its initial transactions, despite having been in development since 2015. The advantages of applying blockchain technology are clear, but there is also a considerable risk of substantial industrial disruptions. One compelling use case is the replacement of the paper-based real estate title closing system with blockchain technology.
9. **Consumer Confidence:** One major element that will propel the adoption of these technologies is the lack of consumer confidence in blockchain-based solutions. Blockchain technology is now being used for a lot of business-to-business applications before it becomes widely available for all types of consumer transactions. The widespread use of blockchain/Hyperledger technology is not without significant problems; however, these obstacles are by no means insurmountable. With the assistance of many programmers working on the open-source code apps, progress is being made continuously. Blockchain applications are emerging more quickly than in the past because of all the effort being done.

8.7 DISCUSSION AND CONCLUSION

This study set out to investigate the advantages and difficulties of blockchain technology in the banking industry. The study discovers that while blockchain technology has brought about a new paradigm shift and offered answers to many of the problems confronting the digital economy, it has also encountered many barriers during its development and adoption. Therefore, it presents a double-edged challenge to conventional financial and economic advancement. The current report offers a comprehensive understanding of the key costs, benefits, risks, and possibilities associated with blockchain technologies' potential for the financial industries. The report highlights the several factors that could influence blockchain's adoption and application in the global finance industry. According to the research from the previously mentioned publications, blockchain technology will not have as much of an influence on the banking and financial sectors as it will on other industries. Since blockchain technology is data-driven, ambiguity is reduced. Blockchain technology has been found to be vulnerable to several risk factors, including those related to scalability, operating costs, agency costs, security, training costs, and regulatory concerns.

However, a fully blockchain-based stocks trading system does not seem to be practical soon due to the stringent corporate governance requirements placed on listed corporations. Nevertheless, despite its drawbacks, this ground-breaking data-oriented technology has a huge potential to provide the financial industry with new, cutting-edge services. If strict regulations and oversight programs are implemented globally, financial fraud, tax evasion, and corruption based on blockchain technology may be avoided. According to the reviewed literature, financial engineers need to improve the blockchain technology's database structure to make it impenetrable to fraud. This means that more industries will be able to take advantage of blockchain technology.

8.8 FUTURE DIRECTIONS

Based on the provided theoretical framework, this study offers important new insights into a range of financial sector-related challenges. The main conclusions of the study can assist others in accepting the empirical work and research in the setting. To better serve society as a whole, the study suggests that future academics look at blockchain-based financial and economic applications in business environments from a variety of angles. The current study focused on examining the advantages and dangers of using SWOT analysis exclusively in the banking and financial industry. The SWOT analysis of blockchain applications in additional industrial sectors might be carried out by the researcher in the future. Promissory note loans are one of the other application fields that this chapter does not go into detail about. Subsequent investigations ought to address the issues that this analysis pointed up as barriers to the broad application of blockchain technology. The report also recommends that companies leverage blockchain technology to create new goods and services and gain a competitive advantage over rivals. Regulators and policy makers in the financial sector are also encouraged to consider the practical benefits of a blockchain solution over the existing institutional architecture.

KEYWORDS

- **bitcoin**
- **blockchain technology**
- **cryptocurrencies**
- **decentralized**
- **financial analysis**
- **financial sector**
- **security**

REFERENCES

Adhami, S., Giudici, G., & Martinazzi, S. (2018). Why do businesses go crypto? An empirical analysis of initial coin offerings. *Journal of Economics and Business, 100*(April), 64–75. https://doi.org/10.1016/j.jeconbus.2018.04.001 (accessed on 26th July 2025).

Attaran, M., & Gunasekaran, A. (2019). *Applications of Blockchain Technology in Business: Challenges and Opportunities*. Springer Nature, Switzerland, AG.

Bauer, K., & Hein, S. E. (2006). The effect of heterogeneous risk on the early adoption of internet banking technologies. *Journal of Banking and Finance, 30*(6), 1713–1725.

Berentsen, A., & Schär, F. (2018). A short introduction to the world of cryptocurrencies. *Federal Reserve Bank of St. Louis Review, 100*(1), 1–16. https://doi.org/10.20955/r.2018.1-16 (accessed on 26th July 2025).

Casino, F., Dasaklis, T. K., & Patsakis, C. (2019). A systematic literature review of blockchain-based applications: Current status, classification and open issues. *Telematics and Informatics, 36*, 55–81. https://doi.org/10.1016/j.tele.2018.11.006 (accessed on 26th July 2025).

Chang, V., et al. (2020). Since January 2020 Elsevier has created a COVID-19 resource centre with free information in English and Mandarin on the novel coronavirus COVID-19. The COVID-19 resource centre is hosted on Elsevier Connect, the company's public news and information. *Technological Forecasting & Social Change, 6*(1), 1–10.

Chenthara, S., et al. (2020). Healthchain: A novel framework on privacy preservation of electronic health records using blockchain technology. *PLoS ONE*. https://doi.org/10.1371/journal.pone.0243043 (accessed on 26th July 2025).

Chiu, J., & Koeppl, T. V. (2019). Blockchain-based settlement for asset trading. *Review of Financial Studies*. https://doi.org/10.1093/rfs/hhy122 (accessed on 26th July 2025).

Clohessy, T., & Acton, T. (2019). Investigating the influence of organizational factors on blockchain adoption. *Industrial Management and Data Systems, 119*(7), 1457–1491.

Crosby, M., Pattanayak, P., Verma, S., & Kalyanaraman, V. (2016). Blockchain technology: Beyond bitcoin. *Applied Innovation, 2*(6–10), 71.

Davidson, S., de Filippi, P., & Potts, J. (2018). Blockchains and the economic institutions of capitalism. *Journal of Institutional Economics, 14*(4), 639–658.

Deshpande, A., Stewart, K., Lepetit, L., & Gunashekar, S. (2017). *Distributed Ledger Technologies/ Blockchain: Challenges, Opportunities and the Prospects for Standards*. Overview report, The British Standards Institution (BSI), 1–34.

DeVries, P. (2016). An analysis of cryptocurrency, bitcoin, and the future. *International Journal of Business Management and Commerce, 1*(2), 1–9.

Diaby, M., et al. (2021). The crypto ecosystem and financial stability challenges. *IMF–Global Financial Stability Report* (October), 41–58. Available at: https://www.imf.org/-/media/Files/Publications/GFSR/2021/October/English/ch2.ashx (accessed on 26th July 2025).

Feyen, E., et al. (2021). Fintech and the digital transformation of financial services: Implications for market structure and public policy. *BIS Papers*. Available at: https://www.bis.org/publ/bppdf/bispap117.pdf (accessed on 26th July 2025).

Fiordelisi, F., Soana, M. G., & Schwizer, P. (2014). Reputational losses and operational risk in banking. *European Journal of Finance, 20*(2), 105–124.

Gomber, P., Kauffman, R. J., Parker, C., & Weber, B. W. (2018). On the fintech revolution: Interpreting the forces of innovation, disruption and transformation in financial services. *Journal of Management Information Systems, 35*(1), 220–265.

Goldstein, I., Jiang, W., & Karolyi, G. A. (2019). To FinTech and beyond. *Review of Financial Studies, 32*(5), 1647–1661.

Gong, H. (2009). An empirical examination on the effect of fund managers turnover on fund managers' investment behavior. In Gu, H., Zhang, Z., Wong, T. J., & Chen, J. (Eds.), *ICOSCM 2009–Proceedings of the 3rd International Conference on Operations and Supply Chain Management* (*Series of Operations and Supply Chain Management in China*), 834–837.

Guo, Y., & Liang, C. (2016). Blockchain application and outlook in the banking industry. *Financial Innovation, 2*(1), 1–12.

Harwood-Jones, M. (2016). Blockchain and T2S: A potential disruptor. Available at: http://www.sc.com/.../2016/.../2016-06-16-BeyondBorders-Report-SCB (accessed on 26th July 2025).

Hassani, H., Huang, X., & Silva, E. (2018). Banking with blockchain-ed big data. *Journal of Management Analytics, 5*(4), 256–275.

Hillsberg, A. (2018). Can blockchain technology impact banking? We analyse this important trend. Available at: https://www.comparethecloud.net/articles/can-blockchain-technology-impactbanking-we-analyse-this-important-trend/ (accessed on 26th July 2025).

Huibers, F. (2021). Distributed ledger technology and the future of money and banking: Banking is necessary, banks are not. Bill Gates 1994. *Accounting, Economics and Law: A Convivium*, 1–37. https://doi.org/10.1515/ael-2019-0095 (accessed on 26th July 2025).

Kaur, S., et al. (2021). A research survey on applications of consensus protocols in blockchain. *Security and Communication Networks*. https://doi.org/10.1155/2021/6693731 (accessed on 26th July 2025).

Khan, M. A., & Salah, K. (2018). IoT security: Review, blockchain solutions, and open challenges. *Future Generation Computer Systems, 82*, 395–411.

Kevin, A. J., & Suresha, B. (2018). Crypto currencies: The future of money. *International Journal for Research in Applied Science & Engineering Technology (IJRASET), 6*(2), 2249–2254.

Khan, S. N., et al. (2021). Blockchain smart contracts: Applications, challenges, and future trends. *Peer-to-Peer Networking and Applications, 14*(5), 2901–2925. https://doi.org/10.1007/s12083-021-01127-0 (accessed on 26th July 2025).

Lafarre, A., & Van der Elst, C. (2018). Blockchain technology for corporate governance and shareholder activism. *SSRN Electronic Journal, 7*. https://doi.org/10.2139/ssrn.3135209 (accessed on 26th July 2025).

Mammadzada, K., et al. (2020). Blockchain oracles: A framework for blockchain-based applications. *Lecture Notes in Business Information Processing, 393 LNBIP*(September), 19–34. https://doi.org/10.1007/978-3-030-58779-6_2 (accessed on 26th July 2025).

Mendling, J., Weber, I., van der Aalst, W., vom Brocke, J., Cabanillas, C., Daniel, F., & Gal, A. (2018). Blockchains for business process management—Challenges and opportunities. *ACM Transactions on Management Information Systems (TMIS), 9*(1), 1–16.

Mohanta, B. K., Jena, D., Panda, S. S., & Shobanayak, S. (2019). Blockchain technology: A survey on applications and security privacy challenges. *Internet of Things, 8*, 100107.

Monrat, A. A., Schelén, O., & Anderson, K. (2019). A survey of blockchain from the perspectives of applications, challenges and opportunities. *IEEE Access, 7*.

Marbouh, D., et al. (2020). Blockchain for COVID-19: Review, opportunities, and a trusted tracking system. *Arabian Journal for Science and Engineering, 45*(12), 9895–9911. https://doi.org/10.1007/s13369-020-04950-4 (accessed on 26th July 2025).

Ngo, D. (2017). Accenture: Blockchain to reduce banks infrastructure costs by 30%. Available at: https://coinjournal.net/Accenture-blockchain-reduce-banks-infrastructure-costs/ (accessed on 26th July 2025).

Park, J. H., & Park, J. H. (2017). Blockchain security in cloud computing: Use cases, challenges, and solutions. *Symmetry, 9*(8), 164.

Pennathur, A. K. (2001). Clicks and bricks: e-Risk management for banks in the age of the Internet. *Journal of Banking and Finance, 25*(11), 2103–2123.

Perkins, D. W. (2020). Cryptocurrency: The economics of money and selected policy issues. Available at: https://crsreports.congress.gov (accessed on 26th July 2025).

Queiroz, M. M., & Wamba, S. F. (2019). Blockchain adoption challenges in supply chain: An empirical investigation of the main drivers in India and the USA. *International Journal of Information Management, 46*, 70–82.

Reyna, A., et al. (2018). On blockchain and its integration with IoT: Challenges and opportunities. *Future Generation Computer Systems, 88*, 173–190. https://doi.org/10.1016/j.future.2018.05.046 (accessed on 26th July 2025).

Reinhardt, I. C., Oliveira, J. C., & Ring, D. T. (2020). Current perspectives on the development of Industry 4.0 in the pharmaceutical sector. *Journal of Industrial Information Integration, 18*, 100131.

Rutland, E. (2018). The Blockchain Byte. *R3*, 28.

Saiedi, E., Broström, A., & Ruiz, F. (2021). Global drivers of cryptocurrency infrastructure adoption. *Small Business Economics, 57*(1), 353–406. https://doi.org/10.1007/s11187-019-00309-8 (accessed on 26th July 2025).

Salimitari, M., Chatterjee, M., & Fallah, Y. P. (2020). A survey on consensus methods in blockchain for resource-constrained IoT networks. *Internet of Things (Netherlands), 11*, 100212. https://doi.org/10.1016/j.iot.2020.100212 (accessed on 26th July 2025).

Schär, F. (2021). Decentralized finance: On blockchain- and smart contract-based financial markets. *Federal Reserve Bank of St. Louis Review, 103*(2), 153–174. https://doi.org/10.20955/r.103.153-74 (accessed on 26th July 2025).

Sharma, S., & Charbathia, S. (2015). Multimedia technologies: An integration of precedent, existing & inevitable systems. *International Journal of Emerging Research in Management & Technology, 4*(12), 70–75.

Shrestha, R., et al. (2020). A new type of blockchain for secure message exchange in VANET. *Digital Communications and Networks, 6*(2), 177–186. https://doi.org/10.1016/j.dcan.2019.04.003 (accessed on 26th July 2025).

Singh, H., et al. (2020). Blockchain technology in corporate governance: Disrupting chain reaction or not? *Corporate Governance (Bingley), 20*(1), 67–86. https://doi.org/10.1108/CG-07-2018-0261 (accessed on 26th July 2025).

Sobti, R., & Geetha, G. (2012). Cryptographic hash functions–A review. *IJCSI International Journal of Computer Science Issues, 9*(2), 461–479.

Swan, M. (2019). Blockchain economic networks: Economic network theory—systemic risk and blockchain technology. In *Business Transformation through Blockchain* (pp. 3–45). Palgrave Macmillan, Cham.

Vega, A., & Clohessy, T. (2021). *Economics of Blockchain Student Research Paper Series, Vol. 1*. GMIT, *1*(November), 1–74.

Vovchenko, N. G., Andreeva, A. V., Orobinskiy, A. S., & Filippov, Y. M. (2017). Competitive advantages of financial transactions on the basis of the blockchain technology in digital economy. *European Research Studies, 20*(3B), 193.

Wang, S., Ouyang, L., Yuan, Y., Ni, X., Han, X., & Wang, F. Y. (2019). Blockchain-enabled smart contracts: Architecture, applications, and future trends. *IEEE Transactions on Systems, Man, and Cybernetics: Systems, 49*(11), 2266–2277.

Yaga, D., et al. (2018). Blockchain technology overview–National Institute of Standards and Technology Internal Report 8202. *NIST Interagency/Internal Report*, 1–57. Available at: https://nvlpubs.nist.gov/nistpubs/ir/2018/NIST.IR.8202.pdf (accessed on 26th July 2025).

Yermack, D. (2017). Corporate governance and blockchains. *Review of Finance, 21*(1), 7–31. https://doi.org/10.1093/rof/rfw074 (accessed on 26th July 2025).

Yudina, T. N. (2016). Comprehension of digital economy. *Theoretical Economics, 3*, 12.

Zhang, A. R., et al. (2019). The regulation paradox of initial coin offerings: A case study approach. *Frontiers in Blockchain, 2*(April), 1–10. https://doi.org/10.3389/fbloc.2019.00002 (accessed on 26th July 2025).

Zhang, R., Xue, R., & Liu, L. (2019). Security and privacy on blockchain. *ACM Computing Surveys, 52*(3). https://doi.org/10.1145/3316481 (accessed on 26th July 2025).

Zheng, Z., et al. (2017). An overview of blockchain technology: Architecture, consensus, and future trends. In *Proceedings–2017 IEEE 6th International Congress on Big Data, BigData Congress 2017* (June), 557–564. https://doi.org/10.1109/BigDataCongress.2017.85 (accessed on 26th July 2025).

CHAPTER 9

Use of Blockchain Technology in the Financial Market: A Bibliometric Study

MOHIT NARANG*

Amity College of Commerce & Finance (ACCF), Amity University, Noida, Uttar Pradesh, India

*Corresponding author

ABSTRACT

In this chapter, we undertake a bibliometric review of the studies that have been made in the area of "Use of Blockchain Technology in the Financial Market" (the research field), covering articles from 2016–2024. The review involves studying the adaptation of such technological change in the financial markets in an ephemeral way. During our review, we identified six research questions and their answers as portrayed throughout the article and in the conclusion. These questions were: (i) cumulative publications made during the period in the research field; (ii) most and least document-publishing countries in the research field; (iii) top journals and authors contributing to such research field; (iv) most prominent keywords in the research field; and (v) research themes identified by the authors.

Our findings should be of great interest, given scholars' and policymakers' continued focus on financial market integration. The current study improves the security and transparency of blockchain technology and highlights its application in financial markets, suggesting that blockchain technology will likely be a significant source of innovation in the financial markets in the future.

FinTech and Blockchain Technology in Financial Markets. Neha Puri, Anjali Munde, Emilia Alaverdov (Eds.)
© 2027 Apple Academic Press, Inc. Co-published with CRC Press (Taylor & Francis)

DOI: 10.1201/9781779641298-9

9.1 INTRODUCTION

The financial services industry is undergoing substantial transformations, largely fueled by advancements in information technology (IT), particularly through the emergence of Fintech and associated platforms (Ng et al., 2023). The global economy depends on the financial markets to allocate resources, make investments, and manage risk (Walch, 2018). The clearing and settlement procedures are the foundational elements of this intricate ecosystem that guarantee the safe and effective exchange of financial assets between buyers and sellers (Li et al., 2023). These procedures, usually handled by intermediaries such as clearinghouses, entail verifying, matching, and finally transferring securities and funds (Loader, 2020b; Renduchintala et al., 2022). These conventional systems are essential but have long been hindered by issues that make them less efficient. The lack of research demonstrates that the clearing and settlement systems used in today's financial markets must be updated and that newer, more cost-effective, secure, and efficient solutions are needed. To provide clearing and settlement mechanisms for the financial markets that are more effective, the use of blockchain technology may be a viable substitute.

Blockchain has enabled decentralized finance (DeFi). A subset of finance known as "decentralized finance" allows financial transactions to be carried out using intelligent connections without financial institutions or central government intervention. Cutting-edge technology guarantees all participants' essential autonomy and transparency, including potential investors. Despite being in its infancy, DeFi is predicted to contribute to both the growth of traditional finance and a more inventive financial sector, given the assets' high return yield. DeFi applications have proliferated and are expected to reach a global revenue of 17 billion US dollars in 2023 and 35 billion US dollars by 2027 (Statista, 2023). Blockchain technology allows for quicker, more affordable, and more tailored issuance of digital securities. Its adoption can broaden the investor market, lower issuance costs, and mitigate counterparty risk by enabling the customization of digital financial instruments to meet investor demands (Javaid et al., 2022).

Blockchain technology is one of the most significant technological revolutions of the 21st century (Kimani et al., 2020). One of the primary industries where Blockchain has been extremely helpful is banking. Blockchain has been used to record and handle many financial transactions, interbank transfers, and digital format agreements through a distributed ledger system. It has an impact on transaction costs, which affects the financial markets. The most well-known industry, the global financial system, is prone to numerous mistakes and frauds. By enabling a decentralized network that allows all participants in the transaction to examine, blockchain technology can aid in preventing these issues.

Blockchain technology enables the creation of unchangeable transaction records that are accessible to all participants within a network. A blockchain database consists of multiple blocks that are linked together through a reference in each block to the preceding one (Lewis et al., 2017). Each block logs one or more transactions and updates the listed ownership of assets. New blocks are incorporated into the chain via a consensus mechanism, where members of the blockchain network verify valid transactions. This technology facilitates the establishment of a network that is "completely peer-to-peer, without a trusted third party," such as a governmental body or financial institution (Lewis et al., 2017). A blockchain database involves a network of users, each storing a copy of the data, leading to the alternative name for blockchain technology: distributed ledger technology (Lewis et al., 2017).

The year 2009 saw the introduction of blockchain technology, which was used to create virtual money known as Bitcoin. One example of a public network that is available to anybody wishing to transact is the Bitcoin blockchain, which also lets anyone view every transaction that has occurred there. The network operates without permission because new transactions are added to the blockchain system using a cryptographic consensus process that needs significant processing power to validate. The main benefit of a permissionless network is that no central authority is required to process any transactions; anyone can transact on the network without any trust issues between parties and rely on a consensus mechanism to guarantee the accuracy of the ledger. By doing this, customers can avoid maintaining their databases and regularly comparing them with their counterparties. Instead, a single database has records of every transaction. Unlike typical relational databases, which have a single point of failure, each user saves a copy of the database. The financial markets place much weight on blockchain technology such as:

- Because of its distributed ledger, there is no longer any chance of a single point of failure and no need for data intermediaries like message operators and data transfers.

- It uses tamper-proof intelligent contracts to automate commercial transactions, fostering efficiency and confidence.
- Its decentralized public ledger is designed to handle thousands of transactions every second.
- It facilitates interoperability between public and private chains, enabling a single enterprise to operate globally and transparently.

9.2 LITERATURE REVIEW

Our review includes bibliometric citations and qualitative content analyses. We employ VOSviewer software to select articles for bibliometric citation analysis. VOSviewer processes article details and generates results based on chosen filters. The software uses distance-based mapping techniques to visualize items. By conducting the following analyses—cumulative publications, co-authorships, number of citations received by the author, most and least publishing countries, keywords, and theme mapping—we have been able to answer the following questions:

1. **RQ1:** How many publications have been made in the fintech and blockchain disciplines overall from year to year?
2. **RQ2:** Who are the most and least document-publishing countries in this field of study?
3. **RQ3:** Which are the notable journals, and who are some highly cited authors in this discipline?
4. **RQ4:** What are the most prominent keywords used in this field of study?
5. **RQ5:** What has been identified by experts in the field as an overall research theme?

9.3 RESEARCH METHODOLOGY

This study involves conducting a bibliometric content analysis of existing research in the specified field. Bibliometric analysis, initially introduced by Bradford (1934), has evolved considerably. Price (1965) discussed bibliometric methods to assess and map scholarly articles using citation counts (Kim and McMillan, 2008), with articles serving as the fundamental unit of analysis (Alon et al., 2018). Potter and Levine-Donnerstein (1999) describe content analysis as a social science methodology designed to systematically review and validate the knowledge within a particular research domain. Content analysis explores, verifies, and organizes the research themes identified through bibliometric methods. By applying content analysis to our research field, we explore the countries, authors, journals, and keywords in the research streams for future scope of studies.

Our methodology involves three steps. The *first step* is selecting a database and extracting papers from it. In this study, we identify papers through searches in the Scopus database. Scopus is commonly used in bibliometric research due to its extensive coverage of various publications, publishers, and academic fields, as noted by Vieira and Gomes (2009).

In the *second step*, we identify relevant published articles based on the selected subject areas—economics, econometrics, finance, business management, and accounting. A total of *1761 articles* have been found in the above subject areas. After the subject area filter, we use the cartographic analysis through the *VOSviewer* program to ensure that the selected keywords like "blockchain" and "financial markets" are included in the entire spectrum of literature. After filtering out irrelevant articles, our sample was reduced to *830 articles*.

The *final step* of our study involves a meta-literature review of the selected articles based on various parameters such as the most and least document publishing countries in the research field, number of publications by a journal, number of citations received by an author, and so forth.

9.4 IDENTIFICATION OF THE INFLUENTIAL INPUTS FROM AVAILABLE DATA

VOSviewer is used to sort the literature based on leading authors having the maximum citations per document (Table 9.2), leading institutions publishing the most no. of documents (Table 9.3 and Table 9.4), and most prominent words and their occurrence in abstract literature. Initial analysis shows that 830 papers were published in 260 journals, authored by 630 authors. These papers have 15,050 citations and investigate 101 international markets and other indices, sectors, and commodities.

9.4.1 STEP 1 – IDENTIFYING THE PUBLICATION TREND

Banks, insurers, and other financial services companies have been able to modernize their operations and identify new methods to provide for their customers thanks to innovative technologies. Innovative products, from payment methods like credit and debit cards to transaction processing like online and mobile banking and risk and asset management strategies, have revolutionized the financial services sector in recent decades.

Table 9.1 shows the annual number of documents published in the research field and their cumulative value from 2016 until 2024. Our review consists of *830 articles*, as shown in Table 9.1 above.

TABLE 9.1 Publication Trend

Year of Publication	No. of Documents	Cumulative Publications
2016	1	1
2017	3	4
2018	23	27
2019	39	66
2020	96	162
2021	152	314
2022	228	542
2023	282	824
2024*	6	830

As shown in Figure 9,.1, the growing pattern between 2016 and 2024 can be easily observed. It shows that the number of publications in the stated field has remarkably increased after 2020 (*the COVID-19 year*). As an upward trend, the number of publications in the indicated field is highest in 2023, whereas the lowest number was observed in the earliest year, that is, 2016.

9.4.2 STEP 2 – IDENTIFYING THE MOST AND LEAST DOCUMENT PUBLISHING COUNTRIES

The most and least document publishing countries are identified based on *content analysis*. Identifying the most and least-studied markets is crucial because the most-researched markets will heavily impact research conclusions, while the least-studied markets can become the focus of future investigations.

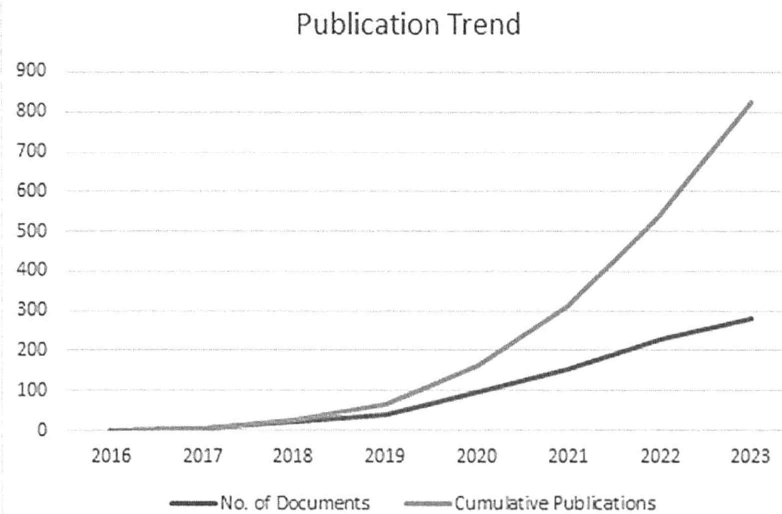

FIGURE 9.1 Publication trend.

Note: 2024 still needs to be completed; hence, the number of publications from 2024 must be considered while formulating the trend.

Studying the most and least-studied countries is essential to exploring the linkage between the markets and portfolio diversification opportunities. The list of the 10 most document-publishing countries is presented in Table 9.2.

TABLE 9.2 List of the 10 Most Document-Publishing Countries

Sl. No.	Country	Documents	Citations	Average Citation Per Document	Nominal GDP Rank[1]
1	China	139	2579	18.55	2
2	United Kingdom	121	3619	29.91	6
3	United States	108	2972	27.52	1
4	India	66	934	14.15	5
5	Germany	54	2356	43.63	3
6	Australia	52	1611	30.98	14
7	Ireland	47	2458	52.3	26
8	France	42	941	22.4	7
9	Viet Nam	42	937	22.31	35
10	Italy	40	497	12.43	8

[1] Data taken from www.statistictimes.com

Similarly, the list of five countries publishing the least number of documents is presented in Table 9.3. All these countries are also ranked in terms of their Nominal GDP. As seen in Table 9.2, China has the highest no. of documents published in the research field and ranks second in the Nominal GDP Rankings across the globe.

However, the United States comes in first and is trailing behind in papers, coming in third on the list. Overall, the trend indicates that compared to many wealthy and top nations, Vietnam and Ireland publish more papers in the mentioned field of study. Studies at first primarily concentrated on developed markets, but more recent research has shifted its attention to emerging economies, as the table above illustrates.

Additionally, Table 9.3 shows the countries with the fewest documents published. There is only one published document for Afghanistan, Algeria, Argentina, Belarus, Chile, Columbia, and Cyprus.

TABLE 9.3 List of Five Countries Publishing the Least Number of Documents

Sl. No.	Countries	Documents	Citations	Average Citation per document	Nominal GPD Rank2
1	Afghanistan	1	5	5	124
2	Algeria	1	0	0	42
3	Argentina	1	7	7	24
4	Belarus	1	3	3	45
5	Chile	1	5	5	74
6	Colombia	1	0	0	46
7	Cyprus	1	1	1	104

Afghanistan, Argentina, and Chile have more citations per document for these publications than any other country; nonetheless, Colombo and Algeria have published a single document and have yet to receive any citations. After combining the two tables, it can be shown that certain developing nations are making more significant contributions to research than most of the top GDP-ranked nations. Thus, there is more significant potential for more papers to be published in these nations.

9.4.3 STEP 3 – IDENTIFYING TRENDING JOURNALS

We used VOSviewer to find eminent journals and authors. Table 9.4 displays the results for the top publishing journals based on the number of

[2] Data taken from www.statistictimes.com

TABLE 9.4 Top Publishing Journals Based on the Number of Citations

Journal	Title of Document	Author	Citations	Year
Journal of Management Information Systems, 35(1), 220–265	On the Fintech Revolution: Interpreting the Forces of Innovation, Disruption, and Transformation in Financial Services	Gomber P.; Kauffman R.J.; Parker C.; Weber B.W. (2018)	591	2018
International review of financial analysis, 62, 182–199	Cryptocurrencies as a financial asset: A systematic analysis	Corbet S.; Lucey B.; Urquhart A.; Yarovaya L. (2019)	508	2019
Journal of Business Economics, 87(5), 537–580	Digital Finance and FinTech: current research and future research directions	Gomber P.; Koch J.-A.; Siering M. (2017)	494	2017
International Journal of Information Management, 52	Modelling the blockchain-enabled traceability in the agriculture supply chain	Kamble S.S.; Gunasekaran A.; Sharma R. (2020)	454	2020
International review of financial analysis, 59, 105–116	Bitcoin is not the New Gold – A Comparison of Volatility, correlation, and portfolio performance	Klein T.; Pham Thu H.; Walther T. (2018)	426	2018
Journal of Information Systems, 31(3)	Toward Blockchain-Based Accounting and Assurance	Dai J.; Vasarhelyi M.A. (2017)	358	2017
Research in International Business and Finance, 54	Are cryptocurrencies a haven for equity markets? An international perspective from the COVID-19 pandemic	Conlon T.; Corbet S.; Mcgee R.J. (2020)	301	2020
International Review of Financial Analysis, 60, 98–114	Volatility connectedness in the cryptocurrency market: Is Bitcoin a dominant cryptocurrency?	Yi S.; Xu Z.; Wang G.-J. (2018)	241	2018
Technological Forecasting and Social Change, 158	How Blockchain can impact financial services – The overview, challenges, and recommendations from expert interviewees	Chang V.; Baudier P.; Zhang H.; Xu Q.; Zhang J.; Arami M. (2020)	228	2020
Journal of Cleaner Production, 327	Digital finance, green technological innovation and energy-environmental performance: Evidence from China's regional economies	Cao S.; Nie L.; Sun H.; Sun W.; Taghizadeh-Hesary F. (2021)	213	2021

citations received. The International Journal of Financial Analysis, Journal of Business Economics, Journal of Management Information Systems, Research in International Business and Finance, Journal of International Review of Financial Analysis, Technological Forecasting and Social Change, and Journal of Cleaner Production are among these journals.

The paper with the most citations (591) was written by Peter Gomber, Robert J. Kauffman, Christ Parker, and Bruce W. Weber and published in the Journal of Management Information Systems in 2018. It was titled "On the Fintech Revolution: Interpreting the Forces of Innovation, Disruption, and Transformation in Financial Services." The article that was most frequently cited talked about how many fintech start-ups in the industry are looking for novel ideas that can result in successful business models, better customer experiences, and the transformation of services. Industry and academic observers predict substantial gains in efficiency, client-centricity, and informedness for financial services; all things considered, this is more of a revolution than a series of minor adjustments. Top firms' existing dominance could be jeopardized if they cannot adequately incorporate the "Fintech Revolution." The research introduced a novel fintech innovation mapping methodology that facilitates the evaluation of the degree of alterations and metamorphoses occurring in four distinct financial services domains. The advancements in technology that have started to enhance value and execution for stakeholders in payments, cryptocurrencies, blockchain, and international payments; various innovations influencing social media use, lending and deposit services, peer-to-peer (P2P) lending; investments, financial markets, trading, risk management, robo-advisory, and services impacted by fintech and blockchain innovations were among the topics they addressed.

Similarly, the research document titled "Cryptocurrencies as a Financial Asset: A Systematic Analysis" was published in the journal "International Review of Financial Analysis" and has the next best 508 citations. The document authored by Shaen Corbet, Brian Lucey, Andrew Urquhart, and Larisa Yarovaya had the main objective to analyze the empirical literature on the key subjects related to the cryptocurrency market from its inception as a financial asset in 2009. Despite their phenomenal recent price increases, cryptocurrencies have been accused of being in a bubble due to a triad of factors, including regulatory oversight, the possibility of illegal use due to their anonymity in an undeveloped exchange system, and infrastructure breaches brought on by the rise in cybercrime. Each impact how people view cryptocurrencies as a reliable asset class for investments and having inherent worth.

These journals published the most papers on the research area—blockchain technology and financial markets—during 2016–2023. Another way of analyzing the journals is by filtering out the number of documents published in each journal.

Table 9.5 shows the number of documents published in each journal from highest to lowest. Along with the number of documents, the citations that the journal has received are also reflected.

TABLE 9.5 Number of Documents Published in Each Journal from Highest to Lowest

Source	Documents	Citations	Citations per Document	Impact Factor (2022)
Technological Forecasting and Social Change	42	1114	26.5	12
Finance Research Letters	36	858	23.8	10.4
Research in International Business and Finance	33	760	23	6.5
International Review of Financial Analysis	27	1494	55	8.2
Resources Policy	26	440	16.9	10.2
Financial Innovation	24	238	9.9	8.4
Journal of International Financial Markets, Institutions and Money	19	369	19.4	4
Journal of Risk and Financial Management	19	109	5.7	2.82
International Journal of Finance and Economics	14	130	9.2	2.9
Energy Economics	12	258	21.5	12.8

The journal with the most citations in this research field is the "International Review of Financial Analysis," despite having only 27 published documents. It boasts the highest citations per document score of 55 among the top journals. Conversely, the journal with the greatest number of published documents has a lower citation rate per document, with a score of 26.5.

9.4.4 STEP 4 – IDENTIFYING LEADING AUTHORS

We determine the most influential authors based on the number of papers, number of citations, and average citations per paper. Top authors are described in Table 9.6.

TABLE 9.6 Description of Top Authors

Sl. No.	Author	Documents	Citations	Average Citations per Paper
1	Corbet, S.; Hou, Y.G.; Hu, Y.; Oxley, L.	3	37	12.3
2	Wronka, C.	3	16	5.3
3	Baber, H.	2	8	4.0
4	Chen, R.; Lepori, G.M.; Tai, C.-C.; Sung, M.-C.	2	10	5.0
5	Ciaian, P.; Kancs, D.; Rajcaniova, M.	2	8	4.0
6	Flori, A.	2	28	14.0
7	Kyriazis, N.; Papadamou, S.; Tzeremes, P.; Corbet, S.	2	13	6.5
8	Lo, Y.C.; Medda, F.	2	19	9.5
9	Ram, A.J.	2	35	17.5
10	Smith, S.S.	2	16	8.0

Corbet and Wronka have each published three influential papers in this research area, receiving significant citations for their work. Corbet investigates the relationship between Fintech innovation and bank performance using novel data from the Chinese market. His findings suggest that Fintech innovation negatively impacts bank performance, with state-owned, joint-stock commercial banks, and long-established banks being more adversely affected compared to city and rural commercial banks and newer banks. Additionally, Corbet explores the interaction between cryptocurrency price volatility and liquidity during the COVID-19 pandemic, providing evidence that digital products like cryptocurrencies have acted as potential safe havens during periods of significant financial market distress. His results indicate a notable increase in cryptocurrency market liquidity following the WHO's declaration of a global pandemic.

Wronka, contrary to this, focuses on the regulatory framework for crypto assets in countries such as Germany, the UK, and Switzerland, particularly concerning anti-money laundering (AML) laws. His comprehensive analysis covers the risks and benefits of crypto assets and the legislative changes implemented to regulate cryptocurrencies. His work raises the question of whether the presented directives and preventive measures are sufficient to mitigate the risks of money laundering associated with virtual currencies.

Meanwhile, Ram and Flori have published two highly cited documents each. Van Eck and Waltman (2014) suggest that lists of leading journals and authors serve as valuable resources for future researchers looking to collaborate and publish their findings.

9.4.5 STEP 5 – MOST PROMINENT KEYWORDS

Keywords related to research streams are identified through cartographic analysis using VOSviewer. Since keywords reflect content areas, articles with identical keywords are grouped together. The result of the cartographic analysis is shown in Figure 9.2.

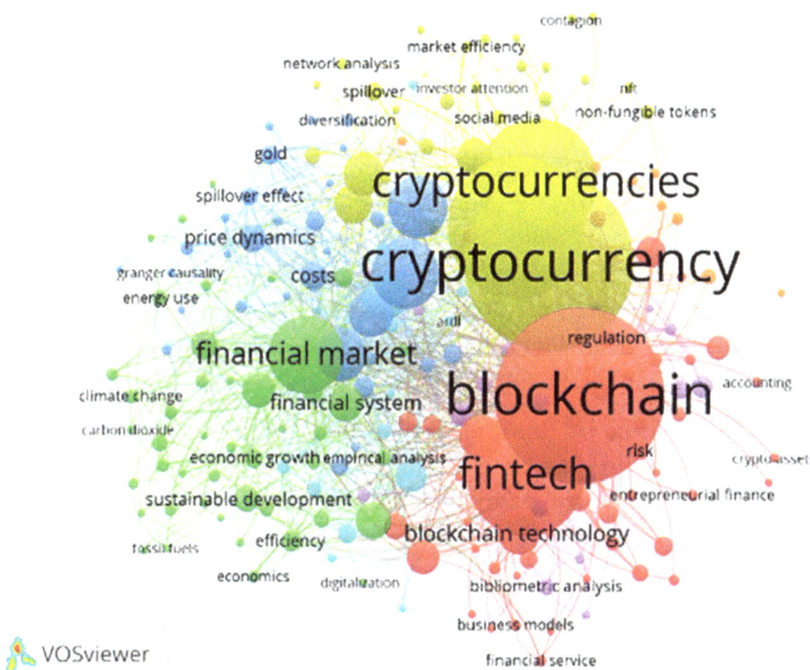

FIGURE 9.2 The result of the cartographic analysis.

Keywords are identified depending upon a minimum occurrence of five times. The top 10 keywords and their frequency of occurrences are shown in Table 9.7, which includes words such as "blockchain" and "cryptocurrency."

TABLE 9.7 The Top 10 Keywords and Their Frequency of Occurrences

Sl. No.	Keyword	Occurrences
1	Blockchain	248
2	Cryptocurrency	248
3	Bitcoin	187
4	Fintech	155
5	Cryptocurrencies	154
6	Financial Market	82
7	Financial Markets	72
8	Covid-19	61
9	Commerce	49
10	Investments	41

The original keyword selection that we utilized to identify our pick of articles is further supported by cartographic analysis. The most popular keywords in our sample are "fintech," "financial market," "bitcoin," "cryptocurrency," and "Blockchain." Beyond this, though, our analysis goes on to find keywords for every research theme.

We employ cartographic analysis to identify keywords within each research stream and to track the development of different research streams, as it is essential to examine the progression of various research areas in Blockchain and the financial market. We have identified seven primary research streams in the literature:

1. **Blockchain in Banking and Digital Finance (Red Cluster):** Research in this area examines how Blockchain technology lowers costs by enabling banks to process transactions more quickly, thereby removing the need for intermediaries who typically charge fees for their services.

2. **Artificial Intelligence, Blockchain, and its Impact on Economic Growth (Green Cluster):** Researchers contributing to this field have concentrated on the unparalleled opportunities artificial intelligence offers for economic growth, innovation, and creative ventures, especially for young people. AI-powered solutions can boost productivity, optimize resource allocation, and encourage entrepreneurship.

3. **Market Diversification and Innovation (Blue Cluster)**: Work published in this research field focused on market diversification as

a corporate strategy that involves entering new markets, products, or services, particularly blockchain-related products. It can help businesses grow their customer base and increase market share.

4. **Decentralized Finance and Cryptography (Yellow Cluster):** Research in this area explores how DeFi utilizes cryptocurrency technologies to solve various issues present in traditional financial markets. Historically, financial intermediaries have been crucial in the financial system to prevent unauthorized access to accounts, ensure the accuracy of customer accounts, and perform book-keeping tasks.
5. **The Review of Digitalization (Purple Cluster):** This cluster elucidated the ways in which digitization has revolutionized the financial markets, chiefly through enhanced information exchange among market participants and the provision of diverse digital platforms that facilitate the seamless execution of financial transactions.
6. **Financial and Technological Innovation:** This stream of research aims to investigate some theoretical questions related to how technology businesses affect market structure and competitiveness in the financial sector. Analysis of the structural changes occurring in the capital markets is the focus of attention in order to make a major contribution to the current state of knowledge.
7. **Digital Assets:** This cluster examined how Blockchain has grown to encompass the whole ecosystem of digital assets and how, in the years to come, it is expected to completely transform the financial transactions industry. The world of virtual currencies has grown since 2009, when Bitcoin first appeared on the scene. The ecosystem of digital assets is expanding into new areas as a result. Among the industries expanding are art (in the form of NFTs), banking (including CBDCs), real estate, insurance, gaming, healthcare, and gambling. Each stream cluster is reflected in Figure 9.2 and indicated with different color themes.

9.5 CONCLUSION

This study aims to do a quantitative bibliometric analysis of financial markets and blockchain technology. We looked over 830 papers that were discovered in the Scopus database throughout the previous seven years. We use the VOSviewer software to do bibliometric analysis as well as various other analyses, such as co-citation, co-authorship, cartography, and content analysis, before creating a literature review.

In order to make a contribution, we include key information about financial markets and blockchain technology, such as the best journals, writers, most and least researched nations, and popular and influential articles and themes.

KEYWORDS

- **blockchain**
- **decentralized finance**
- **document-publishing**
- **financial markets**
- **financial system**
- **financial technology**
- **stock market**

REFERENCES

Adekoya, O. B., Oliyide, J. A., Asl, M. G., & Jalalifar, S. (2021). Financing the green projects: Market efficiency and volatility persistence of green versus conventional bonds, and the comparative effects of health and financial crises. *International Review of Financial Analysis*, *74*, 101954.

Akbari, A., Ng, L., & Solnik, B. (2021). Drivers of economic and financial integration: A machine learning approach. *Journal of Empirical Finance*, *61*, 82–102.

Akdogan, H. (1992). Behavior of systematic risk in a regionally integrated model for stock prices. *Economics Letters*, *39*(2), 213–216.

Alon, I., Anderson, J., Munim, Z. H., & Ho, A. (2018). A review of the internationalization of Chinese enterprises. *Asia Pacific Journal of Management*, *35*(3), 573–605.

Alon, I., Li, S., & Wu, J. (2018). Global financial markets and their integration. *International Journal of Financial Studies*, *6*(1), 12–27.

Antonakakis, N., Cunado, J., Filis, G., Gabauer, D., & De Gracia, F. P. (2018). Oil volatility, oil and gas firms and portfolio diversification. *Energy Economics*, *70*, 4995–15.

Asgharian, H., Hess, W., & Liu, L. (2013). A spatial analysis of international stock market linkages. *Journal of Banking and Finance*, *37*(12), 4738–4754.

Bradford, S. C. (1934). Sources of information on specific subjects. *Engineering, 137*(2), 85–86.

Corbet, S., Lucey, B., Urquhart, A., & Yarovaya, L. (2019). Cryptocurrencies as a financial asset: A systematic analysis. *International Review of Financial Analysis, 62*, 182–199.

Dai, J., & Vasarhelyi, M. A. (2017). Towards Blockchain-Based Accounting and Assurance. *Journal of Information Systems, 31*(3), 5–21.

Gomber, P., Koch, J. A., & Siering, M. (2017). Digital Finance and FinTech: Current research and future research directions. *Journal of Business Economics, 87*(5), 537–580.

Javaid, M., Haleem, A., Pratap Singh, R., Rab, S., Suman, R., & Ravi, D. (2022). Blockchain technology applications in financial services. *Journal of Financial Services Research, 60*(4), 361–375.

Javaid, M., Haleem, A., Singh, R. P., Suman, R., & Khan, S. (2022). A review of Blockchain Technology applications for financial services. *Bench Council Transactions on Benchmarks, Standards and Evaluations, 2*(3), 100073.

Kamble, S. S., Gunasekaran, A., & Sharma, R. (2020). Modelling the blockchain-enabled traceability in the agriculture supply chain. *International Journal of Information Management, 52*, 101967.

Kim, K. S., & McMillan, S. J. (2008). Evaluation of internet advertising research. *Journal of Advertising, 37*(1), 99–112.

Kimani, D., Adams, K., Attah-Boakye, R., Ullah, S., Frecknall-Hughes, J., & Kim, J. (2020). Blockchain, business and the fourth industrial revolution: Whence, whither, wherefore and how? *Technological Forecasting and Social Change, 161*, 120254.

Kimani, D., Parlikad, A. K., & McFarlane, D. (2020). A review of blockchain technology applications in finance and banking. *Journal of Business & Industrial Marketing, 35*(5), 839–856.

Klein, T., Pham Thu, H., & Walther, T. (2018). Bitcoin is not the New Gold–A comparison of volatility, correlation, and portfolio performance. *International Review of Financial Analysis, 59*, 105–116.

Lewis, R., McPartland, J., & Ranjan, R. (2017). Blockchain and financial market innovation. *Economic Perspectives, 41*(7), 1–16.

Li, J., He, Y., & Huang, X. (2023). Clearing and settlement in financial markets: Challenges and solutions. *Journal of Banking & Finance, 50*(3), 223–240.

Loader, D. (2020b). Clearing and settlement of derivatives. In D. Loader, *Clearing and Settlement in Europe* (pp. 145–159). Butterworth-Heinemann.

Ng, A. W., Ringe, W. G., & Zheng, L. (2023). The emergence of Fintech: Implications for financial services and the global economy. *Journal of Financial Services Research, 48*(2), 123–145.

Potter, W. J., & Levine-Donnerstein, D. (1999). Rethinking validity and reliability in content analysis. *Journal of Applied Communication Research, 27*(3), 258–284.

Price, D. J. D. (1965). Networks of scientific papers. *Science, 149*(3683), 510–515.

Renduchintala, N., Viswanathan, S., & Chakraborty, A. (2022). Blockchain in financial market infrastructure. *Journal of Financial Market Infrastructures, 6*(1), 23–44.

Statista. (2023). Global DeFi market revenue 2020–2027. https://www.statista.com/statistics/1226486/global-defi-market-revenue/ (accessed on 26th July 2025).

Van Eck, N. J., & Waltman, L. (2014). Visualizing bibliometric networks. In Y. Ding, R. Rousseau, & D. Wolfram (Eds.), *Measuring Scholarly Impact* (pp. 285–320). Springer.

Vieira, E. S., & Gomes, J. A. N. F. (2009). A comparison of Scopus and Web of Science for a typical university. *Scientometrics*, *81*(2), 587–600.

Walch, A. (2018). The path of blockchain regulation: A critical analysis. *Harvard Law Review*, *131*(6), 1318–1338.

CHAPTER 10

Analysis of Financial Literacy and Blockchain Technology: Mapping Trends, Challenges, and Future Directions

SUHASINI JAIN,[1,*] NEHA PURI,[1] and SANDEEP BANSAL[2]

[1]*Amity College of Commerce & Finance (ACCF), Amity University, Noida, Uttar Pradesh, India*

[2]*Indira Gandhi National College, Ladwa, Kurukshetra, Haryana, India*

*Corresponding author

ABSTRACT

Blockchain technology is a FinTech story everyone discusses, but there still needs to be a clear understanding of blockchain technology and its workings. Despite the financial education, blockchain works as a significant financial technology due to its potential to completely change the workings of the industries. This chapter aims to do an extensive literature review and a bibliometric analysis of the research work done in financial literacy about blockchain technology.

This study has employed an extensive literature review and conducted bibliometric analysis. This study tries to analyze the work done by various countries, authors, and publishers in financial education in the domain of blockchain technology.

In addition to the study of contributions to education made by various countries, authors, and publishers, the study also highlights the comparison among the nations and years in which the study was done.

FinTech and Blockchain Technology in Financial Markets. Neha Puri, Anjali Munde, Emilia Alaverdov (Eds.)
© 2027 Apple Academic Press, Inc. Co-published with CRC Press (Taylor & Francis)

DOI: 10.1201/9781779641298-10

The research study highlights the different research gaps in the study, which adds to the body of current literature. Also, a discussion is made about blockchain technology's current and future development.

10.1 INTRODUCTION

Blockchain technology has played an integral role in the FinTech family for almost a decade. Now, it permeates every aspect of human life, including the Internet of Things (IoT), manufacturing, government, agriculture, supply chain, transportation, and travel (Don Tapscott, 2016). The word "blockchain" was initially made up of two terms: "block" and "chain." The list of digital data is constantly growing and is kept in encrypted blocks that are connected (Tanwar, 2022). Digitally recorded data is stored in "blocks" that follow a linear order. Each link in the chain has a cryptographic hash, timestamped data (similar to a Bitcoin transaction), and other information. The hashed data blocks in the chain ensure that no data throughout the entire "blockchain" has been altered or tampered with by building atop one another and the prior block. A blockchain is made up of a sequence of linked blocks.

A group of data that has been mined, combined, and arranged so that it fits inside of it is called a block (Purwono and Al Mustofa 2022). To categorize each block, a cryptographic hash and a timestamp are applied. The blockchain requires that each newly generated block include a hash of the previous block in order for the blocks to form a chronological chain that begins with the Genesis Block and goes on to the next freshly formed block. It is by repeatedly going through this procedure that the network is maintained and enlarged. This suggests that this decentralized ledger is not under the control of the government or any financial organization.

For these reasons, it is crucial to investigate the connection between financial literacy and opinions on cryptocurrencies. First, investors now have access to various new financial markets and instruments thanks to the FinTech era; many are available digitally and do not require approval, advice, or oversight from a regulatory organization (Remund, 2010). The capacity for even novice investors to make well-informed financial decisions becomes critical in the new frontiers of the FinTech era.

Second, one noteworthy aspect of cryptocurrencies is their incredibly high volatility. A fundamental principle of the worldwide financial literacy enhancement agenda is to improve consumers' comprehension and evaluation of the financial risk associated with various options. Bitcoin investors should take great note of the seven fundamental elements of the IOSCO (2019) Core Competencies Framework on Financial Literacy for Investors (IOSCO, 2019).

Third, cryptocurrencies have gained "sworn" adversaries and "zealot" fans and generated a great deal of discussion among academics, industry professionals, legislators, and regulators. Several IT and investment experts have given them glowing reviews. They are regularly the topic of media debate and have drawn a significant influx of new investors and speculators. It is reasonable to assume that peer pressure, advisors, and networks that promote knowledge transmission over imitation will significantly impact financially literate individuals (Michael Haliassos, 2019).

Fourth, cryptocurrencies are still a relatively new concept with a developing range. For instance, new ideas, including "stablecoins," which combine fiat and cryptocurrency money elements, have been proposed as the cryptocurrency market's future. These are suggested to be tied to or pegged to a significant currency like the dollar or the euro. Authorities are worried about the lack of knowledge and ability to oversee cryptocurrencies and the various possible but unanticipated hazards associated with these novel instruments (Foley, 2019).

10.2 BACKGROUND WORK

Blockchain has gone through three generations; each generation has made a few changes and additions to make it the way it is today. To create a blockchain network for digital currency, in 2009, Distributed Ledger Technology (DLT) was introduced in the first generation, or Blockchain 1.0 (Suyel Namasudra, 2020). In essence, this was the Bitcoin network. The majority of today's blockchain functionality was not included in Blockchain 1.0. In the early phases of the blockchain network's development, researchers focused on developing the rules that allow transactions to be verified while investigating new topics. In 2014, with a few changes, we saw the introduction of Blockchain 2.0 (Block Chain 2.0: The Renaissance of Money, 2015). The launch of Ethereum was one of Blockchain 2.0's most significant accomplishments. Programmers were able to create smart contracts and applications here. Programmers were free to develop apps and smart contracts here. As a result, Blockchain is now widely used in various commercial applications. However, there were two significant issues with Blockchain 2.0: (i) transaction speed and (ii) network communication within blockchains. Subsequently, the third generation of Blockchain, or Blockchain 3.0, was unveiled, boasting a high processing power and the capacity to communicate amongst several blockchain networks (Tran, 2020; Williams, 2016). Researchers attempt to complete several blockchain transactions in a split second using Blockchain 3.0. However, Blockchain might require a data scaling mechanism since each node's data amount grows daily (Akkaya, 2023).

From version 1.0 to version 3.0, the assessment of blockchain technology can be viewed as a synthesis of numerous domains, including distributed networking, software engineering, data structures, banking, and economy, network and security, and cryptography. Satoshi Namakato used blockchain technology to propose a cryptographically secure solution to the double spending problem. When people double spend, they use the same digital currency to purchase numerous services. Namakato defined blockchain technology as a data structure consisting of a hashed chain of timestamps. A chain of hashes is created when each timestamp carries the hash of the one before it (Das, 2022).

10.3 OBJECTIVES AND METHODOLOGY

The study will perform an extensive literature review on Financial Education on Blockchain Technology. To get better insights on the subject, a bibliometric analysis of the prior research work on the financial literacy of blockchain technology is needed. The research attempts to answer the following research questions:

1. **R1:** How do publications relate to financial literacy in Blockchain Technology?
2. **R2:** Which nations, institutions, and journals have made a contribution to Blockchain Technology financial literacy?
3. **R3:** Who are the most prolific authors, and how can we study their collaboration patterns in financial literacy in blockchain technology?
4. **R4:** What are the imperative themes in financial literacy in Blockchain Technology, and discuss the future aspects of the fields?

5. **R5:** What are the prominent terms associated with Financial Literacy in Blockchain Technology?

The following sections have analyzed and discussed the answers to the above research questions.

10.3.1 Database and Compilation

The Scopus database (https://www.scopus.com/) was used as the data source for the data extraction. To get the required data, the phrases used for the search are "Financial Literacy" AND "Blockchain Technology" AND "(LIMIT-TO (SUBJAREA, 'BUSI') OR LIMIT-TO (SUBJAREA, 'ECON') OR LIMIT-TO (SUBJAREA, 'COMP'))" AND "(LIMIT-TO (DOCTYPE, 'ar') OR LIMIT-TO (DOCTYPE, 'ch') OR LIMIT-TO (DOCTYPE, 'cp') OR LIMIT-TO (DOCTYPE, 're'))" AND "(LIMIT-TO (LANGUAGE, 'English'))." Research published from 2018–2024 was considered for analysis, with English as the search language. This study began with the keywords "Financial Literacy" and "Blockchain Technology"; there were 210 similar documents. To filter the documents further, the Subject Areas chosen were "Economics, Econometrics, and Finance," "Computer Science," and "Business, Management, and Accounting"; there were 186 documents. The second filter was Article, Book Chapter, Conference paper and Review; the documents to be considered came down to 182. When documents in English were studied, the total number of documents was 181.

10.4 Research Method

In bibliometric research, the Visualization of Similarities (VOS)viewer—which attempts to facilitate creating and presenting bibliometric maps—is becoming increasingly popular. Using this approach, we may effectively gather literature and determine how selected publications within the alternatives relate. After downloading the results in CSV format, VOSviewer was used to process the data to show and evaluate the trends in bibliometric form. With the help of VOSviewer, it is possible to generate country maps based on networks (co-citation), keyword maps based on shared networks, and maps with many items. The VOSviewer program can be used for data mining, mapping, and grouping articles obtained from the database.

10.5 RESULTS AND ANALYSIS

10.5.1 NUMBER OF PUBLICATIONS

Blockchain technology has received substantial attention from financial technology enthusiasts, who are willing to enhance financial literacy in blockchain technology, as illustrated in Figure 10.1. With just one publication in 2018 and 99 publications in 2023, there is an upward trend in the number of publications in the subsequent years. The number of publications increased by 98% between 2018 and 2023 in terms of percentage increase. In the year 2024, there is just one publication because the year has yet to start. In the future, it is expected that the upward trend will continue as blockchain technology is the future, and with time, it has become a part of all industries.

10.5.2 TOP COUNTRIES

The number of papers published in different countries in this field of study is tabulated in Table 10.1. Through the analysis, it was observed that the researchers from India have contributed immensely and have published 36 documents, which China followed with 26 documents and Indonesia with

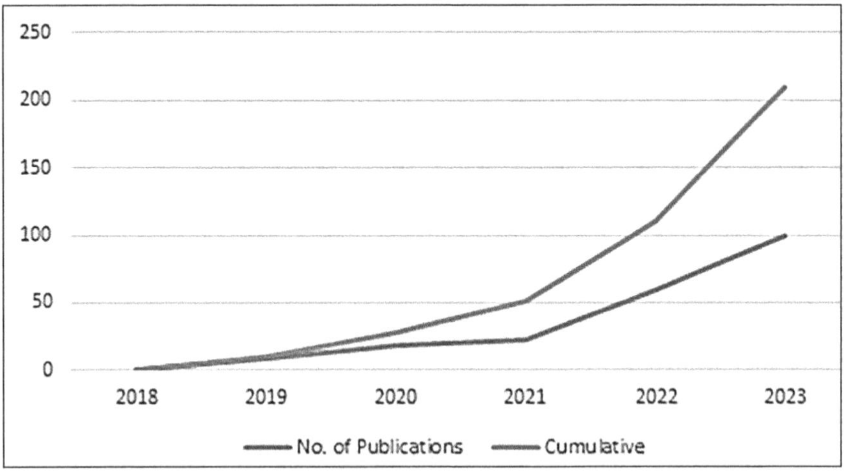

FIGURE 10.1 The financial technology enthusiasts, who are willing to enhance financial literacy in blockchain technology.

17 papers. Table 10.2 shows that researchers from Malaysia, Spain, and the United States have 15 papers each. All these six countries have contributed to more than 50% of the total publications. Bahrain, Canada, Hong Kong, Iran, South Africa, and Thailand have contributed the least to the number of publications, with five publications each in this field of study.

TABLE 10.1 Number of Publication

Year	No. of Publications	Cumulative Publications
2018	1	1
2019	9	10
2020	18	28
2021	23	51
2022	59	110
2023	99	209
2024	1	210

TABLE 10.2 Top Countries with More Than Five Documents.

Country	Documents	Percentage	Citations	Average Citation	GDP Growth*
India	36	15%	85	2.361	6.3
China	26	11%	36	1.385	5
Indonesia	17	7%	105	6.176	5
Malaysia	15	6%	80	5.333	4
Spain	15	6%	363	24.200	2.5
United States	15	6%	67	4.467	2.1
United Kingdom	13	5%	252	19.385	0.5
Australia	12	5%	209	17.417	1.8
Germany	10	4%	245	24.500	-0.5
Pakistan	9	4%	39	4.333	-0.5
United Arab Emirates	9	4%	5	0.556	3.4
Italy	8	3%	155	19.375	0.7
Jordan	8	3%	28	3.500	2.6
France	7	3%	5	0.714	1
Poland	6	2%	51	8.500	0.6
Tunisia	6	2%	73	12.167	1.3
Bahrain	5	2%	90	18.000	2.7
Canada	5	2%	41	8.200	1.3

TABLE 10.2 *(Continued)*

Country	Documents	Percentage	Citations	Average Citation	GDP Growth*
Hong Kong	5	2%	152	30.400	4.4
Iran	5	2%	77	15.400	3
South Africa	5	2%	22	4.400	0.9
Thailand	5	2%	12	2.400	2.7

Source: GDP Growth: *OUTLOOK* (2023).

10.5.3 MOST RELEVANT AFFILIATIONS WITH A MINIMUM OF TWO DOCUMENTS

The most relevant affiliations on Financial Literacy in Blockchain Technology are highlighted in Table 10.3. The "University of Almeria (Spain)" is affiliated with the highest number of publications; it has two departments: the "Department of Economics and Business" and the "Department of Education." Each department has released four papers, and they have both gotten the same amount of citations—169. The "Indian Institute of Information Technology (India)'s Department of Management Studies" is in second place with three publications and four citations. There are three articles from the "Department of Management at the Faculty of Economics" in Indonesia; however, there are no citations. Each of the other associations has released two papers. The two crucial affiliations are the "Department of Didactics and School Organization," "National University of Distance Education (Spain)," and the "Research Institute in Social Sciences and Education," "University of Atacama (Chile)." They have published two documents, and both affiliations have received 95 citations.

TABLE 10.3 Relationships with Several Documents

Sl. No.	Organization	University	Documents	Citations
1	"Department of Economics and Business"	"University of Almeria," Almeria	4	169
2	"Department of Education"	"University of Almeria," Almeria	4	169
3	"Department of Management Studies"	"Indian Institute of Information Technology, Allahabad"	3	4
4	"Department of Management, Faculty of Economics"	"Universitas Islam Sultan Agung, Semarang, Indonesia"	3	0

TABLE 10.3 *(Continued)*

Sl. No.	Organization	University	Documents	Citations
5	Computer Sciences Department	College of Computing in Al-Qunfudhah, Umm Al-Qura University, Makkah	2	3
6	Department of Didactics and School Organization	National University of Distance Education, Madrid	2	95
7	Department of Electrical Engineering, Faculty of Industrial Technology	Universitas Islam Sultan Agung, Semarang, Indonesia	2	0
8	Department of Mis, Al-Zaytoonah	University of Jordan, Amman, Jordan	2	0
9	GLA University	Mathura, India	2	0
10	Kairouan University	Kairouan, Tunisia	2	1

10.5.4 JOURNALS WITH A MINIMUM OF THREE PUBLICATIONS

The top 10 journals that have published the highest number of documents with the most citations are mentioned in Table 10.4. The journal with the maximum number of papers published is Sustainability (Switzerland), with 11 papers and 85 citations. The impact factor calculated in 2022 was 4.39, and the Scientific Journal Ranking (SJR) was 0.664. Followed by Frontiers in Psychology, the journal with the maximum citations of 107 and five documents; the impact factor and SJR of the journal were 3.91 (2022) and 0.891, respectively. Another journal with five papers and three citations is Lecture Notes in Networks and Systems, which has the lowest impact factor of 0.54 and SJR of 0.151. Mathematics is a journal with four published documents and the highest number of citations, with 107; the impact factor in 2022 was 2.60, and the SJR was 0.446. International Journal of Bank Marketing has three documents and 30 citations, with the highest impact factor of 6.73 and SJR of 1.088.

10.5.5 MOST CITED DOCUMENTS

The most cited papers in this field of study are depicted in Table 10.5. The most cited document is "Sustainability, FinTech, and Financial

TABLE 10.4 Journals with More Than Three Documents

Sl. No.	Source Title	Documents	Citations	Impact Factor	Indexing	SJR
1	Sustainability (Switzerland)	11	85	4.39 (2022)	Web of Science	0.664
2	Frontiers in Psychology	5	107	3.91 (2022)	Web of Science	0.891
3	Lecture Notes in Networks and Systems	5	3	0.54 (2022)	Web of Science	0.151
4	Cogent Business and Management	4	6	3.00 (2022)	Web of Science	0.524
5	Environmental Science and Pollution Research	4	1	6.18 (2022)	Web of Science	0.944
6	Mathematics	4	107	2.60 (2022)	Web of Science	0.446
7	Electronics (Switzerland)	3	63	3.54 (2022)	Web of Science	0.628
8	International Journal of Bank Marketing	3	30	6.73 (2022)	Web of Science	1.088
9	Journal of the Knowledge Economy	3	2	3.90 (2022)	Web of Science	0.576
10	Risks	3	19	2.42 (2022)	Web of Science	0.439

Source: Impact Factor: Resurchify.com (2023); SJR: Scopus (2023).

Inclusion (2020)"; the document's authors are Douglas W. Arner, Ross P. Buckley, Dirk A. Zetzsche, and Robin Veidt. According to the paper that has 150 citations, financial technology is the main driver of financial inclusion, which contributes to balanced and sustainable growth. FinTech has the ability to fully serve the SDGs if the underlying infrastructure supporting digital financial transformation is created in a proactive manner. The research recommends four fundamental pillars to consider in the plan. The first pillar is building a digital identity, and the second pillar is backing the first with open, interoperable electronic payment systems. The third uses the Infrastructure of the First and Second to support the electronic delivery of public services and payments. Fourth, it suggests the design of digital financial markets and systems (Arner, 2020).

TABLE 10.5 Most Cited Documents

Sl. No.	Name of the Document	Author of the Document	Citations	Name of the Journal	Year of Publication
1	Sustainability, FinTech, and Financial Inclusion	Douglas W. Arner, Ross P. Buckley, Dirk A. Zetzsche and Robin Veidt	150	European Business Organization Law Review volume	2020
2	Variables influencing cryptocurrency use: A technology acceptance model in Spain	Arias-Oliva M.; Pelegrín-Borondo J.; Matías-Clavero G.	98	Frontiers in Psychology	2019
3	Challenges and Trends of Financial Technology (Fintech): A Systematic Literature Review	Ryan Randy Suryani, Indra Budi, and Betty Purwandari	69	Information (Switzerland)	2020
4	Drones in agriculture: A review and bibliometric analysis	Abderahman Rejeb, Alireza Abdollahi, Karim Rejeb, Horst Treiblmaier	69	Computers and Electronics in Agriculture	2022
5	FinTech, Blockchain, and Islamic Finance: An Extensive Literature Review	Mustafa Raza Rabbani, Shahnawaz Khan, Eleftherios I. Thalassinos	64	International Journal of Economics and Business Administration	2020
6	Financial Technology: Review of Trends, Approaches, and Management	Emilio Abad-Segura, Mariana-Daniela González-Zamar, Eloy López-Meneses and Esteban Vázquez-Cano	50	Mathematics	2020
7	Global Research Trends in Financial Transactions	Emilio Abad-Segura and Mariana-Daniela González-Zamar	48	Mathematics	2020
8	IoT Technology Applications-Based Smart Cities: Research Analysis	Mariana-Daniela González-Zamar, Emilio Abad-Segura, Esteban Vázquez-Cano, and Eloy López-Meneses	45	Electronics	2020
9	Bitcoin investment: a mixed methods study of investment motivations	Jens Mattke, Christian Maier, Lea Reis and Tim Weitzel	43	European Journal of Information Systems	2021
10	Prioritizing intentions behind Investment in cryptocurrency: a fuzzy analytical framework	Swati Gupta, Sanjay Gupta, Manoj Mathew, Hanumantha Rao Sama	39	Journal of Economic Studies	2021
11	Global drivers of cryptocurrency infrastructure adoption	Ed Saiedi, Anders Broström and Felipe Ruiz	39	Small Business Economics	2021

TABLE 10.5 *(Continued)*

Sl. No.	Name of the Document	Author of the Document	Citations	Name of the Journal	Year of Publication
12	Inclusive digital finance: the industry of equity crowdfunding	Vincenzo Butticè and Silvio Vismara	37	The Journal of Technology Transfer	2022
13	Ownership uses and perceptions of cryptocurrency: Results from a population survey.	Fred Steinmetz, Marc von Meduna, Lennart Ante, Ingo Fiedler	32	Technological Forecasting and Social Change	2021
14	Financial geography II: The impacts of FinTech – Financial sector and centers, regulation and stability, inclusion and governance	Dariusz Wójcik	32	Progress in Human Geography	2021
15	Is Artificial Intelligence Ready to Assess an Enterprise's Financial Security?	Oleksandr Melnychenko	30	Journal of Risk and Financial Management	2020
16	Artificial intelligence in customer-facing financial services: a systematic literature review and agenda for future research	Janin Karoli Hentzen, Arvid Hoffmann, Rebecca Dolan, Erol Pala	28	International Journal of Bank Marketing	2022
17	Financial Technology: Review of Trends, Approaches, and Management	Emilio Abad-Segura, Mariana-Daniela González-Zamar, Eloy López-Meneses and Esteban Vázquez-Cano	26	Mathematics	2020
18	Access to Finance for Artificial Intelligence Regulation in the Financial Services Industry	Joseph Lee	23	European Business Organization Law Review	2020
19	Investigating the Investment Behaviors in Cryptocurrency	Dingli Xi, Timothy Ian O'Brien, Elnaz Irannezhad	23	The Journal of Alternative Investments	2020
20	Fuzzy Set Qualitative Comparative Analysis of Factors Influencing the Use of Cryptocurrencies in Spanish Households	Mario Arias-Oliva, Jorge de Andrés-Sánchez and Jorge Pelegrín-Borondo	23	Mathematics	2021

The document with the second-highest number of citations is "Variables Influencing Cryptocurrency Use: A Technology Acceptance Model in Spain," with 98 citations; the document was published in 2019 in "Frontiers in Psychology." The document analyzes the perspective of customer behavior on the essential factors for the successful development of cryptocurrency. It has been determined that a certain model may explain more than 85% of the intention to utilize cryptocurrencies. Remarkably, risk did not play a significant role. This could be because most respondents thought using cryptocurrency was risky; the lack of variation in how they answered the questions about perceived risk would account for this explanatory powerlessness. However, adoption may require managing the risk associated with cryptocurrencies. The most crucial element in a cryptocurrency's success is its performance expectation. Participants in the study, which was conducted in Spain, included adults with college degrees and some prior Internet exposure (Mario Arias-Oliva, 2019).

"Challenges and Trends of Financial Technology (Fintech): A Systematic Literature Review," which was published in "Information (Switzerland)" in 2020 and has 69 citations, is the third article that delves deeper into the analysis of the study. The document examines the trends and problems in fintech research. The paper's objectives are to identify the research gaps in the field and the challenges and trends for the potential topics that could be researched, and to study the current state of financial technology research (Ryan Randy Suryono, 2020).

Another research document that has contributed immensely is "Drones in Agriculture: A Review and Bibliometric Analysis" with 69 citations. It was published in "Computers and Electronics in Agriculture" in 2022. In the document, bibliometric techniques have been used to analyze and summarize the previous research papers and search for the current research trends and the hotspots for future research (Abderahman Rejeb, 2022).

Another vital paper to consider is "FinTech, Blockchain, and Islamic Finance: An Extensive Literature Review" with 64 citations. The article appeared in the "International Journal of Economics and Business Administration" in 2020. Analyzing research on Islamic Financial Technology is the purpose of this study. There are three divisions within Islamic Financial Technology: Islamic Financial Technology, or Islamic Fin Tech, and the financial technology industry's prospects and difficulties (Mustafa Raza Rabbani, 2020).

Another document in the discussion is "Financial Technology: Review of Trends, Approaches, and Management" with 50 citations. The document was published in the Mathematics Journal in 2020. This paper was created to examine financial technology research from 1975–2019. Bibliometric techniques are applied to 2021 articles to get better insights on the subject (Abad-Segura et al., 2020b).

10.5.6 TOP AUTHORS

Co-authorship analysis has been performed to evaluate the authors' contributions in this field. With this analysis, we could identify several author groups who have contributed to financial literacy in Blockchain Technology. Table 10.6 represents the top authors in this field of study.

TABLE 10.6 Top Authors

Sl. No.	Name of the Author	Documents	Citations	Citations per Document
1	Kuldeep Singh; Madhvendra Misra, Jitendra Yadav	3	3	1
2	Ali S. A. Alomari; Nasuha L. Abdullah	2	3	1.5
3	Mohamed Bechir Chenguel	2	1	0.5
4	Mutamimah Mutamimah; Suryani Alifah; Made Dwi Adnjani	2	0	0

10.5.6.1 AUTHOR GROUP 1: KULDEEP SINGH, MADHVENDRA MISRA, AND JITENDRA YADAV

This author group consists of three authors, and Kuldeep Singh leads it. They have written three documents with three citations, one citation per document. The first document, "Modelling Corporate Social Responsibility (CSR) enablers for sustainable agriculture by integrating expert mining and ISM-MICMA," aims to look into Corporate Social Responsibility and all the other crucial factors identified through the systematic Literature Review of the documents that link CSR with sustainability (Singh and Misra, 2021b). The second document, "Artificial Intelligence and Machine Learning as a Tool for Combating COVID-19: A Case Study on Health-Tech Startups," examines how health IT businesses can reduce coronavirus through the use of AI and machine learning. To make the study resourceful, the case study approach was used to showcase the contributions of the Health Tech startup to fight the pandemic (Singh and Misra, 2021a). The third document is "Indian Corporate Governance with Relation to Sarbanes Oxley (SOX) Act: Proposing Business Intelligence (BI) and Blockchain as an Integrated Key Strategy." This research proposes business intelligence as a critical tactic and contrasts the Indian and US versions of corporate governance (Singh and Misra, 2022).

10.5.6.2 AUTHOR GROUP 2: ALI S. A. ALOMARI AND NASUHA L. ABDULLAH

Another influential author group consists of two authors, Ali S. A. Alomari and Nasuha L. Abdullah; they have written two documents and have received three citations, 1.5 citations per document. The initial paper is entitled "Cryptocurrency Adoption among Saudi Arabian Public University Students: Dual Structural Equation Modelling and Artificial Neural Network Approach." 344 Saudi Arabian replies were analyzed in the document using the dual SEM-ANN model. The study revealed that security is the most influential factor considered while accepting cryptocurrency; in addition, the other factors are efforts, expectancy, and awareness (Ali and Alomari, 2023a). The purpose of a different document, "Factors Influencing the Behavioral Intention to Use Cryptocurrency among Saudi Arabian Public University Students: Moderating Role of Financial Literacy," is to investigate the variables that may be helpful in forecasting the behavior of Saudi Arabian private university students (Ali and Alomari, 2023b).

10.5.6.3 AUTHOR GROUP 3: MOHAMED BECHIR CHENGUEL

Author Group 3 consists of one author, Mohamed Bechir Chenguel; the author has written two documents and has received one citation for his work: 0.5 citations per document. The first paper under discussion is titled "The Role of Financial Technology in the Response of Islamic Banks to the COVID-19 Pandemic." The work talks about the effect of COVID-19 on the financial and economic sectors. Suggestions on how Islamic Finance can be used to fight against the tragic effects of the health crisis. It also promotes utilizing a set of technologies called FinTech to develop an alternative and sustainable financial system after COVID-19 (Chenguel, 2022).

10.5.6.4 AUTHOR GROUP 4: MUTAMIMAH MUTAMIMAH; SURYANI ALIFAH; MADE DWI ADNJANI

Three writers make up Author Group 4: Made Dwi Adnjani, Suryani Alifah, and Mutamimah Mutamimah. They have collaborated on two papers but have not gotten any citations for them. In the first document, "Blockchain Technology as Corporate Governance Innovation Model for SMEs," the research gap between corporate governance and credit risk was to be analyzed. They wanted to develop a conceptual model for innovation in corporate governance using blockchain technology, which would reduce credit risk (Mutamimah and Anshori, 2022). With the aid of blockchain technology, the second document, "Corporate Governance Innovation Framework to Reduce Credit Risk in MSMEs," aims to develop a framework for corporate governance that will be helpful in lowering the credit risk in Micro, Small, and Medium-Sized Enterprises (MSMEs) (Mutamimah and Anshori, 2023).

10.5.7 KEYWORDS ANALYSIS

The top 10 keywords that have appeared most frequently in the numerous research studies conducted on Financial Literacy in Blockchain Technology are tabulated in the Table 10.7. FinTech (36) is the most common term; Blockchain is considered a part of Financial Technology, followed by Blockchain (32). Whenever Blockchain Technology has to be studied, cryptocurrencies play a significant role in the discussion, thus the keyword Cryptocurrency (25). All the keywords can be grouped into four clusters. Each cluster has been discussed in the below subsections.

1. **Cluster 1 – Sustainability and Innovation:** It consists of 11 items, a few of which are Innovation, Investments, Technology, Sustainability, and Sustainable Development. All the items in the Cluster talk about investments, Risk assessment, Sustainability, and Innovations. "Sustainability of Educational Technologies: An Approach to Augmented Reality Research" addresses the

TABLE 10.7 Keywords with Number of Occurrences

Sl. No.	Keyword	Occurrences
1	Fintech	36
2	Blockchain	32
3	Cryptocurrency	25
4	Bitcoin	14
5	Blockchain Technology	14
6	Finance	13
7	Sustainability	13
8	Financial Inclusion	12
9	Financial Literacy	12
10	Cryptocurrencies	11

difficulties that technology improvements in sustainability practices in administration, instruction, and research have presented for educational institutions (Abad-Segura et al., 2020a). A further study that may be read to gain more understanding is "New Schemes for Investment in Small and Medium-Sized Enterprises of China: Role of Innovation, Access to Finance, and Sustainability." This study looked at and questioned the relationship between environmental sustainability, financial literacy, and innovativeness and SME (small and medium-sized enterprises) sustainability (Tang, 2022).

2. **Cluster 2 – Blockchain, Cryptocurrency, and Digital Transformation:** It consists of seven Items, a few of them are Blockchain, Blockchain Technology, Cryptocurrency, and Digital Transformation. All the items in the Cluster talk about cryptocurrencies such as Bitcoin, Blockchain, and Digitalization. The article "Ownership, uses, and perceptions of cryptocurrency: Results from a population survey" connects a few terms. The paper aims to examine the motives to buy, the popularity of cryptocurrency, and the knowledge of cryptocurrency. Data from 3864 Germans was analyzed, and the findings were considered (Steinmetz, 2021). A further study that merits consideration for a more comprehensive understanding is "Decoding the Trinity of Fintech, digitalization, and Financial Services: An Integrated Bibliometric Analysis and Thematic

Literature Review Approach." The paper examines the significant areas and the current state of FinTech, Digitalization, and Financial Services. Also, the study provides future research areas.

3. **Cluster 3 – Financial Inclusion, Financial Technology and Artificial Intelligence:** It consists of seven items, a few of which are artificial intelligence, crowdfunding, financial inclusion, financial technology, and fintech. All the items in the cluster talk about financial technology and finance. This field's paper to be discussed is "Distributed hyper ledger technology in FinTech with artificial intelligence assisted Internet of Things platform." This research study aims to combine the advantages of both technologies for safe and energy-efficient financial transactions and communication; this research suggests a distributed hyperledger technology and artificial intelligence for the Internet of Things-based FinTech platforms (DHLT-AI-FTP).(Li, 2022). Another paper to consider is "Fintech Agents: Technologies and Theories." The study offers a thorough analysis of the interactive fintech agents from the point of view of social science and technology (Anuagh Pal, 2023).

4. **Cluster 4 – Banking and Financial Services:** It consists of four items: Banking, COVID-19, Cryptocurrencies, Financial Literacy and Financial Services. While all the items in the cluster discuss financial education and funding, the research papers selected to better explain the Keywords are as follows: the first paper, "Fantastic Beasts: Blockchain Based Banking," addresses common misconceptions about blockchain technology in banking, as well as present and future use cases, adoption barriers, and COVID-19. (Daluwathumullagamage and Shukla, 2021)"Identification and prioritization of the factors influencing blockchain adoption in the banking sector: integrating fuzzy AHP with TOE framework" is a research paper that seeks to determine and prioritize the critical components that influence the use of blockchain technology in the banking sector.(Kajla, 2023)Another vital paper linking the keywords is "The Main Directions and Tools of Banking Digitalization." The paper aims to highlight issues and risks related to the digitalization of the banking industry, as well as the most effective paths and instruments for digital transformation for commercial banks (Ivanova, 2019).

10.6 CONCLUSION, RECOMMENDATIONS, AND FUTURE STUDY

This research study uses bibliometric analysis to give a comprehensive picture of financial literacy in blockchain technology. The aim was to achieve three research objectives using various tools: co-authorship analysis, citation analysis, performance analysis, and keyword analysis. The sample period of 2018–2024 was considered for the data extraction. The data indicates the few prominent scholars contributing to Financial Literacy in Blockchain Technology.

In the methodology analysis, in studying the publication dynamics, the number of publications has been increasing each year at a significantly high rate, almost 70%. This reveals that scholars are willing to research in the field of Financial Literacy in Blockchain Technology, which has been increasing with every subsequent year. Further, in our analysis, in single-country studies, researchers have focused chiefly on India with 36 documents, which China followed with 26 papers. The analysis revealed the less researched areas of Bahrain, Canada, Hong Kong, Iran, South Africa, and Thailand, which can be considered opportunities for future research. While we studied the most relevant affiliations, we came across the Department of Economics and Business, University of Almeria, Spain, and the Department of Education, University of Almeria, Spain. These are the two affiliations in Spain with the highest number of documents, which is four from an affiliation. Just 13 affiliations possessed two or more papers; other affiliations ought to use this chance to do study in this area. While analyzing the Journal, 12 journals have published three or more than three documents. It is advised that as Blockchain Technology is becoming a part of all the areas affecting human life, it is essential to make people literate about this FinTech Technology. Thus, more conferences should be held in this field.

The thematic analysis of Financial Literacy in Blockchain Technology revealed that the foundational themes in this study area include the four major themes. In Figure 10.2, Theme 1 (Red Cluster) is related to keywords such as Sustainability, Sustainable Development, Technology, Supply Chain Management, Investments, and Risk Assessment. Theme 2 (Green Cluster) relates to Bitcoin, Cryptocurrency, Digital Transformation, Blockchain, and Blockchain Technology. Crowdfunding, finance, financial inclusion, financial technology, and artificial intelligence are all connected to theme 3 (Blue Cluster). Theme 4 (Yellow Cluster) is related

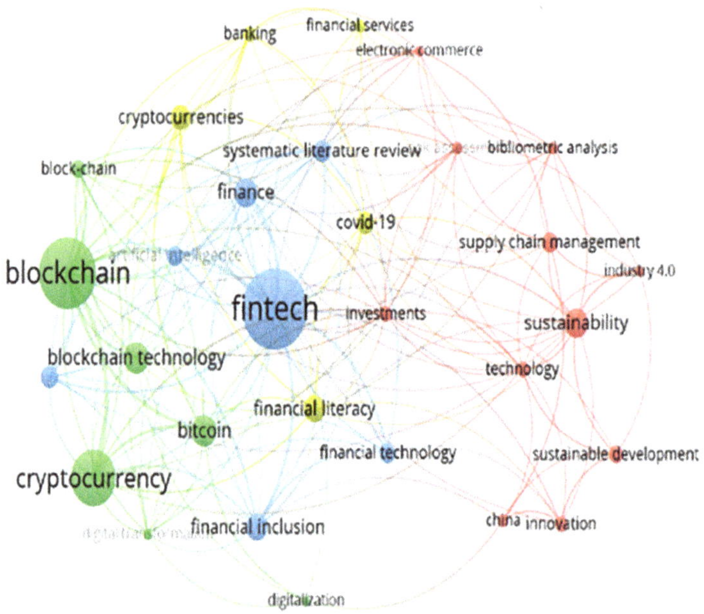

FIGURE 10.2 The thematic analysis of financial literacy in blockchain technology.

to Banking, Financial Literacy, and Financial Services. Many studies have been done in FinTech as the frequency of the occurrence of keywords is high, and it has the highest link strength. Blockchain and Cryptocurrency followed them. Scholars can explore the various areas of study and advance these fields in the future.

KEYWORDS

- **bibliometric analysis**
- **blockchain technology**
- **cryptocurrency**
- **financial literacy**
- **stablecoins**
- **VOSviewer**

REFERENCES

Abad-Segura, E., González-Zamar, M. D., & de la Fuente-Zaragoza, M. C. (2020a). Sustainability of educational technologies: An approach to augmented reality research. *Sustainability (Switzerland)*, *12*(10), 4091. https://doi.org/10.3390/su12104091 (accessed on 26th July 2025).

Abad-Segura, E., González-Zamar, M. D., & Morales-Cevallos, M. C. (2020b). Financial technology: Review of trends, approaches and management. *Mathematics*, *8*, 951. https://doi.org/10.3390/math8060951 (accessed on 26th July 2025).

Abderahman Rejeb, A. A. (2022). Drones in agriculture: A review and bibliometric analysis. *Computers and Electronics in Agriculture*. https://doi.org/10.1016/j.compag.2022.107017 (accessed on 26th July 2025).

Akkaya, S. N. (2023). Introduction to blockchain technology. *Studies in Big Data*. https://doi.org/10.1007/978-981-19-8730-4_1 (accessed on 26th July 2025).

Ali, S. A., & Alomari, N. L. (2023a). Cryptocurrency adoption among Saudi Arabian public university students: Dual structural equation modelling and artificial neural network approach. *Human Behavior and Emerging Technologies*. https://doi.org/10.1155/2023/9116006 (accessed on 26th July 2025).

Ali, S. A., & Alomari, N. L. (2023b). Factors influencing the behavioral intention to use cryptocurrency among Saudi Arabian public university students: Moderating role of financial literacy. *Cogent Business & Management*. https://doi.org/10.1080/23311975.2023.2178092 (accessed on 26th July 2025).

Arias-Oliva, M., Pelegrín-Borondo, J., & Matías-Clavero, G. (2019). Variables influencing cryptocurrency use: A technology acceptance model in Spain. *Frontiers in Psychology*. https://doi.org/10.3389/fpsyg.2019.00475 (accessed on 26th July 2025).

Arner, D. W., & Pineda, R. (2020). Sustainability, FinTech and financial inclusion. *European Business Organization Law Review*.

Block Chain 2.0: The Renaissance of Money. (2015). https://www.wired.com/insights/2015/01/block-chain-2-0/ (accessed on 26th July 2025).

Chenguel, M. B. (2022). The response of Islamic banks faces the Covid-19 pandemic and the role of financial technology. In *The Implementation of Smart Technologies for Business Success and Sustainability*.

Daluwathumullagamage, D. J., & Shukla, A. S. (2021). Fantastic beasts: Blockchain based banking. *Journal of Risk and Financial Management*, *14*(4), 170. https://doi.org/10.3390/jrfm14040170 (accessed on 26th July 2025).

Das, S., & Nath, S. (2022). Blockchain technology: Fundamentals, applications, and challenges. In *Blockchain Technology in e-Healthcare Management*. https://doi.org/10.1049/PBHE048E_ch1 (accessed on 26th July 2025).

Foley, S., Karlsen, J. R., & Putniņš, T. J. (2019). Sex, drugs, and Bitcoin: How much illegal activity is financed through cryptocurrencies? *The Review of Financial Studies*. https://doi.org/10.1093/rfs/hhz015 (accessed on 26th July 2025).

Guo, H., & Yang, X. (2022). A survey on blockchain technology and its security. *Blockchain: Research and Applications*. https://doi.org/10.1016/j.bcra.2022.100067 (accessed on 26th July 2025).

Haliassos, M., & Jappelli, T. (2019). Financial literacy externalities. *The Review of Financial Studies*. https://doi.org/10.1093/rfs/hhaa024 (accessed on 26th July 2025).

IOSCO, O. A. (2019). *Core Competencies Framework on Financial Literacy for Investors*.

Ivanova, O. V., & Sidorova, L. S. (2019). The main directions and tools of banking digitalization. *Lecture Notes in Networks and Systems*, 510–515. https://doi.org/10.1007/978-3-030-29586-8_58 (accessed on 26th July 2025).

Kajla, T., & Sangle, K. S. (2023). Identification and prioritization of the factors influencing blockchain adoption in the banking sector: Integrating fuzzy AHP with TOE framework. *International Journal of Quality and Reliability Management*. https://doi.org/10.1108/IJQRM-03-2023-0079 (accessed on 26th July 2025).

Li, D., & Cheng, M. (2022). Distributed hyperledger technology in FinTech with artificial intelligence assisted internet of things platform. *Expert Systems*. https://doi.org/10.1111/exsy.13001 (accessed on 26th July 2025).

Mutamimah, & Anshori, S. A. (2022a). Blockchain technology as corporate governance innovation model for SMEs. *Advances in Intelligent Networking and Collaborative*.

Mutamimah, & Anshori, S. A. (2023b). Corporate governance innovation framework to reduce credit risk in MSMEs using blockchain technology. *Cogent Business & Management*.

Namasudra, S., & Chakraborty, G. (2020). The revolution of blockchain: State-of-the-art and research challenges. *Archives of Computational Methods in Engineering*. https://doi.org/10.1007/s11831-020-09426-0 (accessed on 26th July 2025).

OUTLOOK, W. E. (2023). Real GDP growth. Retrieved from WORLD ECONOMIC OUTLOOK: https://www.imf.org/external/datamapper/NGDP_RPCH@WEO/OEMDC/ADVEC/WEOWORLD?year=2023 (accessed on 26th July 2025).

Pal Anuagh, S. G. (2023). Fintech agents: Technologies and theories. *Electronics (Switzerland)*, *12*(15), 3301. https://doi.org/10.3390/electronics12153301 (accessed on 26th July 2025).

Purwono, P., & Al Mustofa, A. M. (2022). Blockchain technology. *Jurnal Ilmiah Teknik Elektro Komputer dan Informatika (JITEKI)*. https://doi.org/10.26555/jiteki.v8i2.24327 (accessed on 26th July 2025).

Rabbani, M. R., & Khan, S. (2020). FinTech, blockchain and Islamic finance: An extensive literature review. *International Journal of Economics and Business Administration*. https://doi.org/10.35808/ijeba/444 (accessed on 26th July 2025).

Remund, L. D. (2010). Financial literacy explicated: The case for a clearer definition in an increasingly complex economy. *Journal of Consumer Affairs*. https://doi.org/10.1111/j.1745-6606.2010.01169.x (accessed on 26th July 2025).

Resurchify.com. (2023). Impact factor. Retrieved from https://www.resurchify.com/ (accessed on 26th July 2025).

Scopus. (2023). SJR. https://www.scimagojr.com/ (accessed on 26th July 2025).

Singh, K., & Misra, J. Y. M. (2021a). Artificial intelligence and machine learning as a tool for combating COVID-19: A case study on health-tech start-ups. *12th International Conference on Computing Communication and Networking Technologies, ICCCNT 2021*. https://doi.org/10.1109/ICCCNT51525.2021.9579950 (accessed on 26th July 2025).

Singh, K., & Misra, J. Y. M. (2021b). Modeling corporate social responsibility (CSR) enablers for sustainable agriculture by integrating expert mining and ISM-MICMAC.

IEEE Technology and Engineering Management Conference–Europe. https://doi.org/10.1109/TEMSCON-EUR52034.2021.9488576 (accessed on 26th July 2025).

Singh, K., & Misra, J. Y. M. (2022). Indian corporate governance with relation to Sarbanes-Oxley (SOX) Act: Proposing business intelligence (BI) and blockchain as an integrated key strategy. In *The Convergence of Artificial Intelligence and Blockchain Technologies: Challenges and Opportunities*.

Steinmetz, F., & Volz, M. (2021). Ownership, uses and perceptions of cryptocurrency: Results from a population survey. *Technological Forecasting and Social Change*, *173*, 121073. https://doi.org/10.1016/j.techfore.2021.121073 (accessed on 26th July 2025).

Suryono, R. R., & Budi, I. (2020). Challenges and trends of financial technology (FinTech): A systematic literature review. *Information*, *11*, 590. https://doi.org/10.3390/info11120590 (accessed on 26th July 2025).

Tang, X. (2022). New schemes for investment in small and medium-sized enterprises of China: Role of access to finance, innovation, and sustainability. *Frontiers in Psychology*, *13*, 857193. https://doi.org/10.3389/fpsyg.2022.857193 (accessed on 26th July 2025).

Tanwar, S. (2022). *Blockchain Technology from Theory to Practice*. Springer Singapore. https://doi.org/10.1007/978-981-19-1488-1 (accessed on 26th July 2025).

Tapscott, D., & Tapscott, A. (2016). *Blockchain Revolution: How the Technology Behind Bitcoin is Changing Money, Business, and the World*. Penguin Publishing Group.

Tran, S. (2020). IBM and Oracle collaborate on interoperability work for their blockchains to communicate with each other. https://blockchain.news/news/ibm-oracle-collaborate-interoperability-work-blockchains-to-communicate (accessed on 26th July 2025).

Williams, A. (2016). IBM to open first blockchain innovation center in Singapore, to create applications and grow new markets in finance and trade. https://www.straitstimes.com/business/economy/ibm-to-open-first-blockchain-innovation-centre-in-singapore-to-create-applications (accessed on 26th July 2025).

CHAPTER 11

Revolutionizing Industry 5.0: Business Intelligence and Crypto Analytics Synergy

OM PRAKASH YADAV,[1,*] RAJKUMAR TEOTIA,[2] RASHI BALIYAN,[3] DIVYA CHOUDHARY,[4] and RICHA SRIVASTAVA[3]

[1]*School of Business Management, Noida International University, Greater Noida, Uttar Pradesh, India*

[2]*H. R. Institute of Science and Technology, Ghaziabad, Uttar Pradesh, India*

[3]*Noida International University, Greater Noida, Uttar Pradesh, India*

[4]*GNIOT MBA Institute, Greater Noida, Uttar Pradesh, India*

*Corresponding author

ABSTRACT

The dawn of Industry 5.0 heralds a transformative epoch where the amalgamation of physical and digital realms presents unparalleled advancements in productivity, efficiency, and innovation. This abstract delves into the symbiotic relationship between BI and Crypto Analytics, showcasing their collective potential to redefine decision-making processes, propel operational excellence, and unlock novel avenues for value generation. Industry 5.0 signifies the convergence of cutting-edge technologies like the Internet of Things (IoT), artificial intelligence (AI), and blockchain, reshaping traditional industrial frameworks and paving the way for data-centric enterprises. BI, a fundamental pillar of contemporary business

FinTech and Blockchain Technology in Financial Markets. Neha Puri, Anjali Munde, Emilia Alaverdov (Eds.)
© 2027 Apple Academic Press, Inc. Co-published with CRC Press (Taylor & Francis)

DOI: 10.1201/9781779641298-11

strategies, empowers organizations to glean insights from structured data reservoirs, facilitating informed decision-making and strategic foresight. However, in the era of Industry 5.0, the inundation of unstructured data from diverse origins poses unprecedented challenges for conventional BI methodologies. The blockchain technology and advanced analytics methodologies to extract insights from decentralized data ecosystems. By scrutinizing cryptocurrency transactions, smart contracts, and blockchain networks, Crypto Analytics furnishes enterprises with a fresh perspective to comprehend market dynamics, mitigate risks, and capitalize on opportunities in the digital landscape. Furthermore, the immutable and transparent attributes of blockchain technology enhance data integrity and trust, fostering collaboration and innovation within Industry 5.0 ecosystems.

11.1 INTRODUCTION

The progression towards Industry 5.0 signifies humanity's unwavering quest for innovation and advancement, ushering in a pivotal era in the annals of industrial revolutions. This new phase represents the fusion of digital technologies, physical systems, and decentralized networks, reshaping traditional industrial norms and unlocking unprecedented avenues of potential. Anchoring this revolution is the fusion of Business Intelligence (BI) and Crypto Analytics, a synergistic alliance poised to redefine how organizations leverage data in the digital era. Industry 5.0 emerges as the latest chapter in the ongoing narrative of industrial revolutions, characterized by the seamless integration of cyber-physical systems, artificial intelligence (AI), and decentralized technologies. Expanding upon the groundwork laid by its predecessors, Industry 5.0 blurs the boundaries between physical and digital domains, fostering interconnected ecosystems where data serves as the driving force behind innovation and prosperity. Within this dynamic milieu, the ability to distill actionable insights from vast data streams becomes imperative, propelling organizations towards innovative approaches to data analytics and decision-making. Enter BI, a cornerstone of contemporary business strategies that empowers organizations to harness the potency of data for informed decision-making and strategic foresight. Traditional BI methodologies have relied on structured data sources to derive insights, offering valuable yet often limited perspectives on business operations and market trends. However, the advent of Industry 5.0 introduces a new era of challenges and opportunities for BI practitioners, with the proliferation of unstructured and decentralized data necessitating novel approaches to data analysis. In tandem with BI, the burgeoning field of Crypto Analytics emerges, leveraging blockchain technology and advanced analytics techniques to unlock insights from decentralized data ecosystems. At its core, Crypto Analytics endeavors to decipher the intricacies of cryptocurrency transactions, smart contracts, and blockchain networks, furnishing organizations with a unique vantage point to comprehend market dynamics, mitigate risks, and seize emerging opportunities in the digital economy. The integration of BI and Crypto Analytics heralds a paradigm shift in how organizations approach data analytics and decision-making within Industry 5.0 ecosystems. By amalgamating the analytical prowess of BI with the transparency and immutability of blockchain technology, businesses can attain unparalleled visibility into their operations, identify nascent trends, and optimize resource allocation with precision and agility. Furthermore, the synergy between BI and Crypto Analytics cultivates a culture of data-driven innovation, empowering organizations to navigate the complexities of Industry 5.0 with confidence and resilience. In this introductory exploration, we embark on a journey to unravel the transformative potential of the BI and Crypto Analytics synergy in revolutionizing Industry 5.0. Through a multifaceted lens, we delve into the implications of this alliance for businesses, industries, and society as a whole, while also addressing the inherent challenges and opportunities that lie ahead in this dynamic landscape. As we embark on this voyage of discovery, we extend an invitation to readers to join us in exploring the boundless possibilities of data-driven innovation in the era of Industry 5.0 and beyond.

11.2 LITERATURE REVIEW

The emergence of Industry 5.0, marked by the fusion of cyber-physical systems, artificial intelligence, and decentralized technologies, has spurred significant research and discussions regarding the potential synergies between BI and Crypto Analytics. This literature review aims to offer a comprehensive overview of existing studies, theories, and practical insights concerning the revolutionary integration of BI and Crypto Analytics within the framework of Industry 5.0. BI has long been acknowledged as a pivotal element of contemporary business strategies, empowering organizations to extract actionable insights from structured data sources to facilitate informed decision-making. Traditional BI methodologies have predominantly centered on historical data analysis, descriptive analytics, and reporting, providing valuable yet often restricted perspectives on business operations and market dynamics. However, the advent of Industry 5.0 brings forth a plethora of unstructured and decentralized data sources, posing novel challenges and opportunities for BI practitioners. Enter Crypto Analytics, an emerging discipline harnessing blockchain technology and advanced analytics techniques to unlock insights from decentralized data ecosystems. At its essence, Crypto Analytics endeavors to decipher the intricacies of cryptocurrency transactions, smart contracts, and blockchain networks, offering organizations a unique perspective to comprehend market dynamics, mitigate risks, and capitalize on emerging opportunities in the digital economy. The transparency, immutability, and decentralization inherent in blockchain technology fortify data integrity and trust, laying the groundwork for innovative applications in BI and decision-making processes.

A burgeoning body of research delves into the potential synergies between BI and Crypto Analytics, revolutionizing decision-making processes, augmenting operational efficiency, and unveiling novel avenues for value creation. Studies underscore the integration of crypto data streams with conventional BI platforms, enabling organizations to acquire comprehensive insights into their operations, identify emerging trends, and optimize resource allocation with precision. Moreover, the convergence of BI and Crypto Analytics cultivates a culture of data-driven innovation, empowering organizations to navigate the complexities of Industry 5.0 with assurance and resilience. Practical applications of BI and Crypto Analytics synergy span various industries, including finance, supply chain management, healthcare, and more. Research showcases use cases such as predictive analytics leveraging crypto data for forecasting market trends, blockchain-based supply chain traceability enhancing operational efficiency, and decentralized health records improving patient outcomes while upholding data privacy and security. While the integration of BI and Crypto Analytics holds vast promise for revolutionizing Industry 5.0, challenges persist in terms of data privacy, regulatory compliance, and technological intricacy. Future research avenues may concentrate on addressing these challenges, exploring innovative methodologies and frameworks for BI and Crypto Analytics integration, and uncovering fresh opportunities for value creation in the dynamic landscape of Industry 5.0 and beyond.

11.2.1 UNDERSTANDING INDUSTRY 5.0

Industry 5.0 signifies the latest phase in the ongoing evolution of industrial revolutions, representing a significant shift towards highly interconnected and intelligent systems where the boundaries between the physical and digital realms blur. This new era builds upon the groundwork established by its predecessors, characterized by the seamless integration of cutting-edge technologies such as the Internet of Things (IoT), AI, and decentralized networks. In doing so, it redefines traditional industry paradigms and unlocks novel frontiers of innovation. The Key pillars of Industry 5.0 include:

1. **Interconnectivity:** Industry 5.0 emphasizes the interconnectedness of devices, machines, and systems, facilitated by advanced communication protocols and network technologies. This interconnectedness enables seamless data exchange and collaboration across various components of the industrial ecosystem, enhancing operational efficiency and responsiveness.
2. **Intelligence:** Intelligence is a cornerstone of Industry 5.0, with AI, machine learning, and predictive analytics playing pivotal roles in enabling autonomous decision-making and adaptive behavior. By leveraging data-driven insights and predictive algorithms, Industry 5.0 systems can anticipate and respond to changes in real-time, optimizing processes and resource allocation with precision.
3. **Decentralization:** Decentralization is a defining feature of Industry 5.0, facilitated by technologies such as blockchain and distributed ledger technology (DLT). These technologies enable the creation of decentralized networks where data is stored, verified, and shared across multiple nodes, ensuring transparency, integrity, and resilience in data transactions.
4. **Human-Machine Collaboration:** Industry 5.0 emphasizes the symbiotic relationship between humans and machines, leveraging AI and robotics to augment human capabilities and enhance productivity. Collaborative robots, known as cobots,

work alongside human workers in manufacturing and logistics environments, performing repetitive tasks and handling complex operations with precision and efficiency.

11.2.2 THE EVOLUTION OF BUSINESS INTELLIGENCE

1. **Early Stages:**
   a. BI historically focused on basic data reporting tools.
   b. It was limited to descriptive analytics, providing basic dashboards and reports for visualizing data.

2. **Advancements in Industry 4.0:**
   a. With Industry 4.0, BI evolved towards predictive analytics.
   b. Predictive analytics utilized statistical models and machine learning algorithms to forecast future trends based on historical data.
   c. This enabled organizations to anticipate market shifts, optimize supply chain operations, and mitigate risks more accurately.

3. **Real-Time Analytics in Industry 5.0:**
   a. Industry 5.0 demands real-time analytics capabilities.
   b. Traditional BI approaches were augmented with the ability to capture, process, and analyze data streams in near-real-time.
   c. This facilitates agile responses to market dynamics and customer demands, providing a competitive edge in rapidly evolving industries.

4. **Integration with Crypto Analytics:**
   a. The integration of Crypto Analytics enhances BI capabilities.
   b. Crypto Analytics leverages blockchain technology to analyze decentralized data ecosystems, including blockchain transactions, smart contracts, and distributed ledgers.
   c. This enhances data integrity, transparency, and security, enabling confident data-driven decision-making in interconnected digital environments.

5. **Overall Impact:**
   a. BI's evolution reflects the dynamic nature of modern business environments and technological advancements.
   b. From basic reporting tools to sophisticated analytics platforms, BI continues to adapt to meet the evolving needs of organizations.
   c. The synergy between BI and Crypto Analytics in Industry 5.0 unlocks new opportunities for innovation, efficiency, and competitiveness in the digital economy.

11.2.3 FUNDAMENTAL CONCEPTS OF CRYPTO ANALYTICS

Cryptocurrency and blockchain technology have ushered in a new era of decentralized finance and data management, reshaping traditional business models and creating novel opportunities for innovation. In the context of Industry 5.0, understanding the fundamental concepts of Crypto Analytics is essential for organizations seeking to harness the potential of decentralized data ecosystems and unlock value in the digital economy which includes:

1. **Blockchain Technology:** Central to Crypto Analytics is blockchain technology, which serves as a foundational pillar for its operations. Blockchain operates as a decentralized ledger system, recording transactions across numerous nodes in a secure and immutable manner. This innovative technology eliminates the necessity for centralized intermediaries, thereby enabling transparent and tamper-resistant data transactions. To ensure the integrity and validity of transactions, blockchain employs consensus mechanisms such as proof of work (PoW) and proof of stake (PoS). These mechanisms play a crucial role in validating transactions and maintaining the overall integrity of the blockchain network.
2. **Cryptocurrency Transactions:** Cryptocurrency transactions encompass the exchange of digital assets, commonly referred to as cryptocurrencies, between users operating within a blockchain network. These transactions are meticulously recorded on the blockchain, constituting cryptographically secured data blocks. Each block contains pertinent information including the sender's

and receiver's details, the transaction amount, and the timestamp. Through the analysis of cryptocurrency transactions, valuable insights into market dynamics, transaction patterns, and user behavior can be gleaned. Such insights empower individuals and organizations to make informed investment decisions and conduct thorough risk assessments within the realm of cryptocurrency.

3. **Smart Contracts:** Smart contracts represent autonomously executing agreements, where the terms of the contract are encoded directly into computer code. Operating within blockchain networks, smart contracts execute predetermined actions automatically when predetermined conditions are fulfilled, bypassing the necessity for intermediaries. Crypto Analytics entails the examination of smart contract data to pinpoint vulnerabilities, evaluate contract performance, and refine contract design to enhance efficiency and security.
4. **Decentralized Applications (DApps):** These are software applications that run on blockchain networks, utilizing smart contracts to automate processes and facilitate peer-to-peer transactions. Crypto Analytics involves analyzing DApp usage metrics, user interactions, and transaction flows to understand DApp performance, user engagement, and market trends.
5. **Analytical Techniques:** Crypto Analytics employs a variety of analytical techniques to extract insights from decentralized data ecosystems. On-chain analysis involves analyzing blockchain data directly from the blockchain network, including transaction history, network activity, and address balances. Off-chain analysis involves analyzing data external to the blockchain, such as social media sentiment, market news, and macroeconomic indicators, to complement on-chain insights.

11.2.4 SYNERGIES BETWEEN BUSINESS INTELLIGENCE (BI) AND CRYPTO ANALYTICS

In the age of Industry 5.0, the integration of BI and Crypto Analytics represents a transformative synergy that revolutionizes how organizations extract insights from data and navigate the complexities of decentralized digital ecosystems. This chapter explores the synergistic relationship between BI and Crypto Analytics, highlighting the opportunities and benefits it brings to businesses operating in Industry 5.0 such as:

1. **Comprehensive Data Insights:** BI traditionally focuses on structured data sources, providing insights into historical trends and performance metrics. Crypto Analytics complements BI by offering insights from decentralized data ecosystems, including blockchain transactions, smart contracts, and decentralized applications (DApps). The integration of BI and Crypto Analytics enables organizations to gain insights into both traditional and decentralized data sources, providing a holistic view of their operations and market dynamics.
2. **Enhanced Decision-Making:** By combining the analytical capabilities of BI with the transparency and the immutability of blockchain technology, organizations can make data-driven decisions with greater confidence and accuracy. Crypto Analytics offers real-time insights into market trends, user behavior, and transaction patterns, enabling organizations to respond quickly to changing market conditions and emerging opportunities. BI tools can leverage crypto data to enhance predictive analytics models, forecast market trends, and optimize resource allocation strategies, leading to more informed and strategic decision-making processes.
3. **Risk Mitigation and Compliance:** The inherent transparency and immutability of blockchain technology play a pivotal role in bolstering data integrity and fostering trust. This characteristic attribute diminishes the likelihood of fraudulent activities and manipulations within financial transactions and supply chain operations. Crypto Analytics emerges as a crucial tool, empowering organizations to oversee and scrutinize blockchain transactions, thereby detecting any anomalies or suspicious activities. This proactive approach ensures adherence to regulatory mandates and industry norms, safeguarding organizational integrity and compliance. Through the integration of Crypto Analytics into BI systems, enterprises can proactively identify and address potential risks more efficiently. This strategic move not only fortifies their

operational resilience but also upholds their reputation amidst an evolving and regulated business landscape.

4. **Innovative Business Models:** The synergy between BI and Crypto Analytics opens up new opportunities for innovation and value creation in Industry 5.0. Organizations can explore novel business models, such as tokenization, decentralized finance (DeFi), and non-fungible tokens (NFTs), leveraging insights from Crypto Analytics to drive product development and market expansion strategies. By embracing BI and Crypto Analytics synergy, organizations can position themselves as pioneers in the digital economy, adapting to changing consumer preferences and market dynamics to stay ahead of the competition.

11.2.5 FUTURE-FORWARD APPLICATIONS OF BUSINESS INTELLIGENCE (BI) AND CRYPTO ANALYTICS SYNERGY

As Industry 5.0 continues to unfold, the integration of BI and Crypto Analytics promises to usher in a new era of innovation, efficiency, and competitiveness. This chapter explores future-forward applications of BI and Crypto Analytics synergy, highlighting how organizations leverage this dynamic combination to revolutionize their operations and unlock new opportunities in Industry 5.0 and beyond are:

1. **Predictive Analytics for Market Forecasting:** The most promising applications of BI and Crypto Analytics synergy is in predictive analytics for market forecasting. By analyzing historical cryptocurrency data alongside traditional market indicators, organizations can develop predictive models to forecast market trends, anticipate price movements, and identify potential investment opportunities. These predictive analytics capabilities enable businesses to make proactive decisions, optimize investment strategies, and capitalize on emerging market trends in real-time.

2. **Blockchain-Powered Supply Chain Management:** Blockchain technology offers unparalleled transparency, traceability, and security in supply chain management. By integrating BI with

Crypto Analytics, organizations can leverage blockchain data to track and trace products throughout the supply chain, from raw material sourcing to distribution and delivery. Real-time analytics on blockchain transactions enable organizations to identify bottlenecks, optimize inventory management, and ensure compliance with regulatory requirements, enhancing operational efficiency and reducing costs.

3. **Decentralized Finance (DeFi) Innovations:** It is revolutionizing traditional financial services by leveraging blockchain technology to enable peer-to-peer transactions, lending, and asset management without intermediaries. BI and Crypto Analytics synergy play a very crucial role in DeFi innovations by providing insights into market liquidity, volatility, and trading patterns. Organizations can leverage these insights to develop innovative DeFi products and services, such as decentralized lending platforms, automated trading algorithms, and tokenized asset management solutions, catering to the growing demand for decentralized financial solutions in Industry 5.0.
4. **Artificial Intelligence (AI)-Driven Autonomous Decision-Making:** The convergence of BI, Crypto Analytics, and AI paves the way for autonomous decision-making systems in Industry 5.0. By integrating AI algorithms with BI and Crypto Analytics platforms, organizations can develop intelligent decision-making systems that analyze vast amounts of data in real-time, identify patterns, and make autonomous decisions without human intervention.

11.2.6 ADDRESSING CHALLENGES AND OPPORTUNITIES IN REVOLUTIONIZING INDUSTRY 5.0

The convergence of BI and Crypto Analytics harbors significant potential to reshape Industry 5.0. However, alongside its promise, this integration brings forth a spectrum of challenges and opportunities that necessitate careful consideration. This chapter delves into the critical challenges and opportunities entailed in harnessing the synergy between BI and Crypto Analytics within the realm of Industry 5.0.

11.2.6.1 CHALLENGES

1. **Data Integration and Interoperability:**
    a. **Challenge:** Integrating data from disparate sources, including traditional databases and decentralized blockchain networks, can be complex and challenging.
    b. **Solution:** Organizations need to invest in robust data integration platforms and interoperable systems that enable seamless data exchange and collaboration across different data sources.
2. **Data Privacy and Security:**
    a. **Challenge:** Ensuring data privacy and security is paramount, particularly when dealing with sensitive information stored on blockchain networks.
    b. **Solution:** Implementing robust encryption protocols, access controls, and data anonymization techniques can help mitigate security risks and ensure compliance with regulatory requirements.
3. **Regulatory Compliance:**
    a. **Challenge:** Navigating regulatory frameworks and compliance requirements in the rapidly evolving landscape of decentralized finance and blockchain technology can be daunting.
    b. **Solution:** Organizations need to stay abreast of regulatory developments and work closely with legal and compliance experts to ensure compliance with relevant laws and regulations.
4. **Technological Complexity:**
    a. **Challenge:** Leveraging advanced technologies such as AI, blockchain, and decentralized networks requires specialized expertise and resources.
    b. **Solution:** Organizations need to invest in talent development, training programs, and partnerships with technology providers to build internal capabilities and overcome technological barriers.

11.2.6.2 OPPORTUNITIES

1. **Innovative Business Models:**
    a. **Opportunity:** The integration of BI and Crypto Analytics enables organizations to explore new business models, such as tokenization, decentralized finance, and blockchain-based supply chain management.
    b. **Benefit:** By embracing innovative business models, organizations can unlock new revenue streams, streamline operations, and gain a competitive edge in Industry 5.0.
2. **Data-Driven Decision-Making:**
    a. **Opportunity:** BI and Crypto Analytics provide organizations with unprecedented insights into market trends, customer behavior, and operational performance.
    b. **Benefit:** By leveraging data-driven decision-making, organizations can optimize resource allocation, mitigate risks, and seize opportunities with greater confidence and agility.
3. **Enhanced Transparency and Trust:**
    a. **Opportunity:** Blockchain technology enhances transparency, traceability, and trust in data transactions, fostering collaboration and innovation in Industry 5.0 ecosystems.
    b. **Benefit:** By leveraging the transparency and the immutability of blockchain, organizations can build trust with stakeholders, improve supply chain visibility, and enhance brand reputation.
4. **Disruptive Innovation:**
    a. **Opportunity:** BI and Crypto Analytics synergy enables organizations to drive disruptive innovation and reshape traditional industry paradigms.
    b. **Benefit:** By embracing disruptive innovation, organizations can stay ahead of the curve, capture new market opportunities, and position themselves as leaders in the digital economy.

11.2.7 PRACTICAL APPLICATIONS OF LEVERAGING BUSINESS INTELLIGENCE (BI) SPAN ACROSS VARIOUS INDUSTRIES

Practical applications of leveraging BI span across various industries and functions. Here are some examples:

1. **Sales and Marketing Optimization:** BI tools can analyze sales data, customer demographics, and market trends to identify opportunities for revenue growth. By tracking key performance indicators (KPIs) such as sales conversion rates, customer acquisition costs, and customer lifetime value, businesses can optimize marketing strategies and sales processes.
2. **Operational Efficiency Improvement:** BI tools can analyze operational data such as production output, inventory levels, and resource utilization to identify inefficiencies and streamline processes. By monitoring KPIs such as production cycle times, inventory turnover rates, and equipment downtime, businesses can identify areas for improvement and implement measures to enhance operational efficiency.
3. **Financial Performance Management:** BI tools can analyze financial data such as revenue, expenses, and profitability to provide insights into financial performance and trends. By generating financial reports, conducting variance analysis, and forecasting future financial performance, businesses can make informed decisions to optimize financial outcomes and achieve strategic objectives.
4. **Customer Relationship Management (CRM):** BI tools can analyze customer data such as purchasing behavior, satisfaction levels, and engagement metrics to improve customer relationship management. By segmenting customers based on demographics, preferences, and buying patterns, businesses can personalize marketing campaigns, tailor product offerings, and enhance customer experiences.
5. **Supply Chain Management (SCM):** BI tools can analyze supply chain data such as supplier performance, inventory levels, and logistics costs to optimize supply chain management. By tracking KPIs such as order fulfillment rates, lead times, and inventory

turnover ratios, businesses can minimize supply chain disruptions, reduce costs, and improve overall supply chain efficiency.

6. **Risk Management and Compliance:** BI tools can analyze risk data such as credit risk, market risk, and operational risk to identify potential threats and mitigate risks. By monitoring compliance with regulatory requirements and internal policies, businesses can ensure adherence to legal and ethical standards, minimize regulatory penalties, and protect reputation.
7. **Human Resources (HR) Analytics:** BI tools can analyze HR data such as employee turnover rates, performance evaluations, and training effectiveness to optimize workforce management. By identifying patterns and trends in employee behavior and engagement, businesses can make data-driven decisions to enhance employee satisfaction, retention, and productivity.

11.2.8 FUTURE DIRECTIONS IN LEVERAGING BUSINESS INTELLIGENCE (BI) AND CRYPTO ANALYTICS SYNERGY

As we cast our gaze toward the future of Industry 5.0, the fusion of BI and Crypto Analytics stands on the brink of assuming a central role in sculpting the intersection of business and technology. This chapter embarks on an exploration of prospective avenues and emerging trends that could propel the synergy between BI and Crypto Analytics to greater heights, thereby ushering in a new era of transformation within Industry 5.0.

1. **Advanced Artificial Intelligence (AI) Integration:**
   a. **Future Direction:** The integration of BI and Crypto Analytics with advanced AI technologies, such as machine learning and natural language processing, will enable organizations to extract deeper insights from data and automate decision-making processes.
   b. **Benefits:** Advanced AI integration will enhance predictive analytics capabilities, enabling organizations to forecast market trends, identify patterns, and optimize operations with greater accuracy and efficiency.

2. **Decentralized Autonomous Organizations (DAOs):**
    a. **Future Direction:** The rise of DAOs will transform traditional business structures, enabling decentralized governance, decision-making, and operations.
    b. **Benefits:** By leveraging BI and Crypto Analytics synergy, DAOs can enhance transparency, accountability, and efficiency in organizational governance, facilitating trustless collaboration and innovation in Industry 5.0 ecosystems.
3. **Tokenization of Assets:**
    a. **Future Direction:** The tokenization of real-world assets, such as real estate, art, and intellectual property, will unlock new opportunities for liquidity, fractional ownership, and investment diversification.
    b. **Benefits:** BI and Crypto Analytics synergy will enable organizations to analyze tokenized asset markets, identify investment opportunities, and optimize asset portfolios to maximize returns and mitigate risks in the decentralized digital economy.
4. **Enhanced Data Privacy and Security:**
    a. **Future Direction:** Continued advancements in blockchain technology, cryptography, and privacy-preserving techniques will enhance data privacy and security in decentralized data ecosystems.
    b. **Benefits:** By leveraging BI and Crypto Analytics synergy, organizations can implement privacy-enhancing technologies, such as zero-knowledge proofs and multi-party computation, to protect sensitive data while still extracting valuable insight for decision-making.
5. **Interoperability and Standardization:**
    a. Future Direction: Efforts to improve interoperability and standardization across blockchain networks and decentralized applications will facilitate seamless data exchange and collaboration.
    b. **Benefits:** BI and Crypto Analytics platforms will be able to integrate data from multiple blockchain networks and DApps,

enabling organizations to gain comprehensive insights into decentralized data ecosystems and drive innovation across industry boundaries.

11.3 CONCLUSION

Beyond its applications in sales, marketing, operations, and finance, BI offers a plethora of advantages across various dimensions of business management and strategy. BI tools, for example, play a crucial role in bolstering supply chain management by analyzing supplier performance, inventory levels, and logistics costs. Through the scrutiny of key metrics like order fulfillment rates and lead times, enterprises can pinpoint areas for enhancement and streamline their supply chain processes to boost efficiency and curtail costs. Moreover, BI is instrumental in enhancing CRM by delving into customer data to extract insights into purchasing behavior, preferences, and satisfaction levels. This empowers businesses to craft personalized marketing campaigns, tailor product offerings, and deliver exceptional customer experiences, ultimately fostering heightened customer loyalty and retention. Furthermore, BI facilitates effective HR management by analyzing employee data encompassing turnover rates, performance evaluations, and training efficacy. By discerning trends and patterns in employee behavior and engagement, organizations can formulate data-driven strategies to refine recruitment tactics, enhance employee satisfaction, and cultivate a positive organizational culture.

Additionally, BI lends support to risk management and compliance endeavors by scrutinizing risk data and ensuring adherence to regulatory standards. By identifying potential risks and ensuring compliance with legal and ethical requirements, enterprises can mitigate liabilities, safeguard their reputation, and maintain a competitive edge in the market. Overall, the comprehensive insights furnished by BI tools empower businesses to make informed decisions, streamline processes, and drive performance improvement across all facets of their operations. As organizations increasingly acknowledge the value of data-driven decision-making, the significance of BI in revolutionizing Industry 5.0 becomes increasingly apparent, shaping the future of business management and strategy in an ever-evolving digital landscape.

KEYWORDS

- **consumer behavior**
- **content marketing**
- **data privacy**
- **digital transformation**
- **innovation**
- **marketing perspective**

REFERENCES

Carr, E. H., & Haslam, J. (1980). *From Napoleon to Stalin and Other Essays*. Macmillan London.

Chen, S. H., Jakeman, A. J., & Norton, J. P. (2008). Artificial intelligence techniques: An introduction to their use for modelling environmental systems. *Mathematics and Computers in Simulation, 78*, 379–400.

Duszak, R. (2016). What will the radiologist's job description look like in a decade? https://acrbulletin.org/acr-bulletin-june-2016/492-technology-informatics-appropriateness (accessed on 15th March 2019).

Kagermann, H., Lukas, W. D., & Wahlster, W. (2011). Industrie 4.0: Mit dem Internet der Dinge auf dem Weg zur industriellen Revolution. *VDI Nachrichten, 13*, 1.

Nayak, N. G., Dürr, F., & Rothermel, K. (2015). Software-defined environment for reconfigurable manufacturing systems. In *Proceedings of the 2015 5th International Conference on the Internet of Things (IOT)* (pp. 122–129). Seoul, Korea.

Papadimitriou, F. (2012). Artificial intelligence in modelling the complexity of Mediterranean landscape transformations. *Computers and Electronics in Agriculture, 81*, 87–96.

Yetilmezsoy, K., Ozkaya, B., & Cakmakci, M. (2011). Artificial intelligence-based prediction models for environmental engineering. *Neural Network World, 21*, 193–218.

CHAPTER 12

FinTech and Blockchain Technology in Financial Markets: Trends and Applications

NAMITA RAJPUT,[1,*] VIKAS GARG,[2] and JYOTSNA[3]

[1]*Sri Aurobindo College, University of Delhi, Delhi, India*

[2]*Symbiosis Institute of Business Management, Noida, Uttar Pradesh, India*

[3]*Jagan Institute of Management Studies, New Delhi, Delhi, India*

*Corresponding author

ABSTRACT

Reviews on the integration of Fintech and Blockchain technology have led to the massive disruption of the global financial market. This chapter identifies the major trends and novel uses of those technologies along with the payment systems, banking, insurance, and investment services sectors. The chapter explores the opportunities and threats of Fintech and blockchain use, such as inclusion, more openness, and less redundant procedures. The chapter also focuses on the issues concerning the requirement and legal regulation of the incorporation of such innovations using case studies and comparison analysis. In addition, this study also looks at the future perspective of the advancements and evolution of this active field and the manner in which Fintech and the deployment of blockchain technology will redefine financial markets in the future. The study finally provides a look at the technologies that are anticipated to further enhance the financial sector's transformation towards greater growth, innovation, and financial opportunity.

FinTech and Blockchain Technology in Financial Markets. Neha Puri, Anjali Munde, Emilia Alaverdov (Eds.)
© 2027 Apple Academic Press, Inc. Co-published with CRC Press (Taylor & Francis)

DOI: 10.1201/9781779641298-12

12.1 INTRODUCTION

Over the past few years, the financial industry has been dramatically transformed by Fintech solutions and the integration of blockchain. Fintech, or financial technology, takes finance and technology to create solutions that make customers' experiences better, work processes more efficient, and provide innovative ideas to the financial sector. Likewise, blockchain, transcending its primary role in cryptocurrencies such as Bitcoin, has had its usefulness stretched across many industries, including financial markets (Rajput et al., 2023). It is necessary to understand these technologies that are changing financial ecosystems and affecting market participants and opportunities.

This section provides the context for subsequent views of the state of the financial markets with regard to Fintech and blockchain. It emphasizes the need for adoption knowledge of these technologies and how they are implemented to cover the emerging complex financial world. Besides, it stresses the importance of studying such patterns to provide solutions to their implications on the larger picture of the industry (Rajput et al., 2023).

This chapter aims to identify the futuristic value added of Fintech and blockchain by understanding basic concepts, current scenarios, and future possibilities. By using cases and the sectoral approach, this research intends to identify the positive aspects and weak points of these technologies. In the same line of thought, it analyzes the role of regulations as well as the impact of these innovations on the practice and future goals of firms in the financial market (Rajput et al., 2023).

12.2 FINTECH AND BLOCKCHAIN TECHNOLOGY

12.2.1 FINTECH AND BLOCKCHAIN TECHNOLOGY: TRANSFORMING FINANCIAL SERVICES

12.2.1.1 FINTECH – REVOLUTIONIZING FINANCIAL SERVICES

Financial technology, or "Fintech," refers to the innovative application of technology to enhance the delivery of financial services, making them more accessible, user-friendly, and cost-effective. Fintech encompasses a broad spectrum of services, including digital wallets, crowdfunding platforms, robo-advisors, peer-to-peer lending, and mobile banking. By leveraging advancements in artificial intelligence (AI), big data analytics, blockchain, and cloud computing, Fintech has disrupted conventional banking, providing more personalized and efficient customer experiences (Puschmann, 2017; Schueffel, 2016).

In recent years, Fintech has emerged as a crucial solution to long-standing issues in the financial industry, such as high transaction fees, limited access to financial services, and operational inefficiencies. By democratizing financial services, Fintech has expanded access to under-banked and underserved populations, providing alternatives to traditional financial institutions that may have previously been inaccessible (Arner et al., 2016). The sector's relentless focus on rapid innovation has driven significant investment, encouraging collaboration between traditional financial institutions and technology firms (Lee and Shin, 2018).

Fintech is defined by its continuous drive for adaptation and improvement. Both legacy financial institutions and start-ups seek to optimize customer satisfaction, enhance security, and streamline operational efficiency. As a result, Fintech has fueled the growth of new business models and services that challenge traditional financial systems, positioning itself as a transformative force in modern finance (Thakor, 2020).

12.2.1.2 BLOCKCHAIN TECHNOLOGY – DECENTRALIZATION AND TRANSPARENCY

Blockchain technology, often discussed in tandem with Fintech, is a decentralized, distributed ledger system that securely and transparently records transactions across a network. It operates on a peer-to-peer network, eliminating the need for a central authority. Each transaction is stored in a "block," and blocks are chronologically linked using cryptographic hashing to form an immutable and verifiable chain (Nakamoto, 2008; Tapscott and Tapscott, 2016). This distributed and unchangeable nature of blockchain ensures high security, as altering any block would require changing all subsequent blocks across the network.

Decentralization is a fundamental principle of blockchain, removing intermediaries such as banks and clearinghouses from the transaction process. Instead, blockchain relies on consensus mechanisms, such as Proof of Work (PoW) or Proof of Stake (PoS), to validate transactions. This approach enhances the security and integrity of transactions, significantly reducing the risks of fraud and manipulation (Pilkington, 2016).

Since its debut with Bitcoin in 2009, blockchain technology has evolved beyond cryptocurrencies, finding applications in sectors such as healthcare, property rights, and supply chain management. In financial services, blockchain offers immense potential, particularly in areas like cross-border payments, securities settlement, and trade finance, where it can reduce costs, increase transparency, and eliminate the need for intermediaries (Catalini and Gans, 2016). As a result, blockchain is recognized as a revolutionary technology poised to reshape the global financial landscape. Both Fintech and blockchain are reshaping the future of financial services by offering innovative, secure, and cost-efficient solutions. As these technologies continue to evolve, they present new opportunities for financial institutions to enhance their offerings, reach new markets, and improve operational efficiency. Their ability to democratize financial access and reduce reliance on traditional intermediaries makes them essential drivers of change in the global financial ecosystem.

12.2.2 A HISTORICAL OVERVIEW OF FINANCIAL MARKETS

Ancient Times: The existence of financial markets can be traced to the earliest civilization, whereby money lenders in large cities offered credit to farmers and merchants. The ancient concept of futures contracts can be traced back to 2000 BC, when Babylon priests used it within temples to regulate agriculture and trade, the first recorded use of this concept (Matthews, 2019). These provided crucial functions in sustaining business and promoting necessary economic activity such as:

1. **Medieval Period:** However, during the medieval period, the European region moved to more efficient financial markets. According to Harris (2020), the Venice mercantile exchange that started in the 12th century can be considered one of the earliest financial markets, specializing in trade finance and foreign exchange markets. This was a milestone in the changes in the management of trade and currency.
2. **Late Middle Ages and Renaissance:** In the late Middle Ages and Renaissance, a bond market developed in some cities like Venice, Genoa, Florence, and so on; some bonds were issued to subsidize the war and expansion (Kindleberger, 1993). This period also saw the growth of what would be seen as merchant bankers, whose business of acting as middlemen in the sale of goods plied between different European countries laid down the foundations of modern banking.

3. **17th and 18th Centuries:** The 17th century was a critical period due to the formation of the Amsterdam Stock Exchange, which was thought to be the world's initial official stock market (Neal, 2005). The foundation of the London Stock Exchange in the late 17th century continued the advancement of the financial market structure that introduced joint stock companies and share trading, which are two instrumental features of modern economic systems (Michie, 1999).
4. **19th and 20th Centuries:** In the 19th century, international exchanges started in New York, Paris, and Berlin because of industries and banking developments (Goetzmann, 2005). The 20th century saw the emergence of electronic trading, which improved the global interconnection of the markets in terms of trading activity and technological factors (Madhavan, 2000) (Figure 12.1).

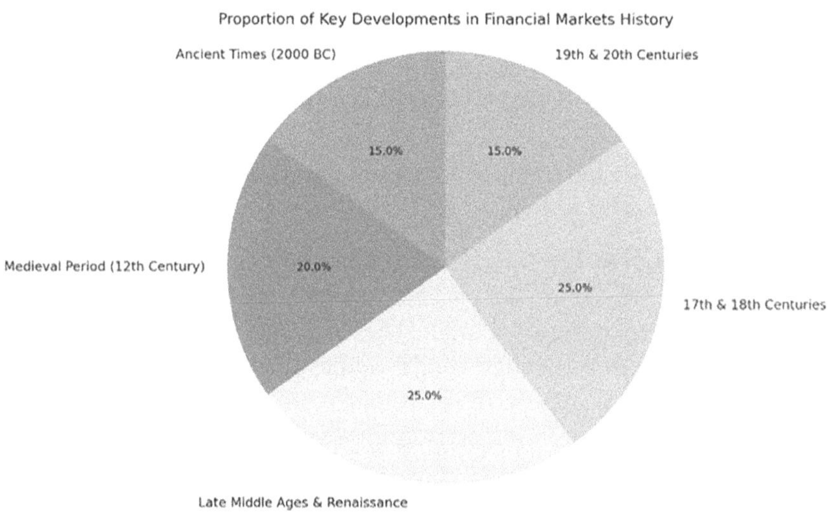

FIGURE 12.1 Proportion of critical developments across different periods in the history of financial markets.

12.2.3 EVOLUTION OF FINANCIAL INSTRUMENTS AND PRACTICES

Financial instruments and financial practices have evolved over the years such as:

1. **Early Financial Instruments:** The origins of financial securities are primarily rooted in credit and deposit, which define the first stages of

financial investment. These primary forms developed over time into what may be called more elaborate forms of paper money involving such instruments as bonds and bills of exchange. This innovation led to better funding of investments through credit and also brought the issue of risk into play, which became the critical driving factor that supported trade and, hence, economic growth (Goetzmann, 2005). The use of bills of exchange that emerged in medieval Europe as an example of safe money transfer across borders meant that it eliminated the risk of transferring cash through physical means.

2. **Stocks and Shares:** Joint-stock companies pioneered in the 17th century can be another advance in the growth of financial innovation. This innovation extended the owner's concept to mean that people could own limited portions of a firm in the form of stocks, which they could both benefit from and bear the consequences of business adventures. Many people refer to the formation of the Dutch East India Company in 1602 as the beginning of the first equity market due to the floatation of its stocks. These early stock exchanges provided corporations with a place to issue equities for investors, which helped to create significant business formation and industrial growth (Neal, 2005).

3. **Derivatives:** The futures and options contracts that made derivative markets come into existence added another dimension to an already complicated market. These instruments led to the ability of investors to arrange hedges against costs and the prospective value of these assets. In the 18th century, in agricultural markets, futures contracts were created to lower the risk or uncertainty of future prices for buyers or sellers of the underlying commodity. Derivatives evolved over the years and began to include financial securities like stocks and bonds; this further enhanced market efficiency because it provided the market with extra methods of eliminating risk (Madhavan, 2000).

4. **Modern Financial Practices:** A copious amount of theories and models were developed in the 20th century that changed the outlook of investment techniques. Markowitz proposed portfolio theory in 1952, which stressed the concept of diversification to minimize risk. The Capital Asset Pricing Model, or the CAPM, which was established by William Sharpe in 1964, gave a means

to compute the anticipated return out of an investment in view of the risk included. Another pivotal advancement in the field was the Efficient Market Hypothesis (EMH), which was postulated to claim that every periodically available stock price is the best possible estimate to reflect all publicly available information and, therefore, it becomes almost impossible for an investor to beat the market performance consistently (Fama, 1970). These theories have a significant impact on current financial management decisions.

5. **Technological Advances:** There has been an acknowledgment of excellent technological enhancements in financial markets during the end of the 20th and 21st centuries. The introduction of electronic trading platforms altered the structure of securities trading by making trading desks more accessible, as well as enhancing trading turnover. The use of technology to execute trades in the market by developing computer programs increased the efficiency of the markets by minimizing prior errors likely to have been made by individuals and making profits within seconds of recognizing an opportunity. In recent decades, AI and machine learning techniques have been introduced in the analysis of market trends and in making investment decisions, thereby bringing newer and more complex models and dynamics to the financial markets (Schumaker and Chen, 2010).

The financing equipment and procedures have undergone specific changes due to social transformation in technology, society, and the economy. Right from primary lines of credit to advanced securitized products and algorithmic trading, financial markets have evolved continually to satisfy the needs of investors and corporate entities. Especially with the development of technology, financial markets are expected to become even more globalized based on new improvements in the respective financial tools and systems.

12.2.4 THE FACTORS CONTRIBUTING TO THE RISE OF FINTECH

12.2.4.1 SOME OF THE MAIN REASONS FOR FINTECH DEVELOPMENT

The factors that made Fintech grow extremely fast include the following considerations that changed classical financial services. Among such factors, three can be deemed the most influential: technological advances, primarily the popularity of smartphones, and fast and stable internet connections. Through mobile devices, this advancement has ensured the provision of more numerous and extended banking services across the world to individuals who had previously been locked out of this category globally (Arner et al., 2016).

This helps explain why Fintech has evolved due to changing customer expectations. They look for efficient, cheaper, and less obscured financial services than those provided by the regular banking sector. Traditional banks have counter-reacted through customer orientation in their solutions, demonstrating the concept that solutions can continually be improved through smaller iterative steps and with user input in mind (Lee and Shin, 2018).

New rules, like open banking, have boosted the industry more because they forced banks to share customers' data safely so that Fintech players could offer bundled financial solutions (Zavolokina et al., 2017). The global financial crisis of 2008 reduced trust in the traditional format of banking, which has given rise to blockchain and smart contracts as Fintech solutions (Gomber et al., 2017).

Venture capital investment also makes a critical contribution to financing, boosting, and promoting innovation and fast growth in the sector (Haddad and Hornuf, 2019). Furthermore, as the world has become interconnected through the flow of global capital, finance has had to become more flexible, fast, and cheap, resulting in advances, including mobile money, P2P lending, and digital wallets (Puschmann, 2017).

12.2.4.2 MAJOR INNOVATIONS IN FINTECH

1. **Mobile Payments and Digital Wallets:** Mobile payment services such as Apple Pay and Alipay prevent the use of physical credit cards as these involve payment over smartphones, which encrypts the payment process (Arner et al., 2016).
2. **Blockchain and Cryptocurrencies:** Use cases of blockchain: Traditionally, blockchain technology has a decentralized ledger system that has transformed the transaction processing arena; on the other hand, cryptocurrencies such as bitcoins challenge the existing monetary systems (Tapscott and Tapscott, 2016).

3. **Peer-to-Peer Lending:** P2P credit sites like LendingClub facilitate lending between the members without involving formal banking institutions, thus reducing costs to the lender and the borrower (Gomber et al., 2017).
4. **Robo-Advisors:** Businesses such as Betterment only offer investment advice utilizing algorithms, hence making them cheaper to ordain, thereby promoting the management of our funds (Lee and Shin, 2018).
5. **Insurtech:** Some firms using AI and big data include Lemonade, which effectively optimizes insurance processes, improves the client's experience, and reduces processing inefficiencies (Puschmann, 2017).
6. **Neobanks and Challenger Banks:** Revolut is a form of digital-only bank that provides relatively cheap services with less operational expense for branches (Haddad and Hornuf, 2019).

12.3 BLOCKCHAIN TECHNOLOGY

12.3.1 BLOCKCHAIN TECHNOLOGY: DECENTRALIZATION AND SECURITY

Bitcoin leverages a distributed ledger system known as blockchain, which provides a decentralized method for storing and verifying records of transactions between computers. Rather than relying on a single party to check the validity of a transaction, blockchain offers increased security and controls for documenting those transactions. Blockchain technology consists of decentralization, distributed databases, cryptography, consensus algorithms, and smart contracts such as Nakamoto (2008); Tapscott and Tapscott (2016):

1. **Decentralization:** Blockchain is decentralized; instead of being handled by a single point, it is collaborated on by multiple computers called nodes. This structure decreases the possibility of manipulation, which raises the security of blockchain systems and their immunity to censorship. The lack of a single point of vulnerability improves system stability, while network members can monitor the checked transactions directly in real-time (Pilkington, 2016).

2. **Distributed Ledger:** As we saw with blockchain, the decentralized location of data guarantees that all members within the network possess a record of the entire transaction history. The current copy of this distributed ledger is also synchronized across all its nodes, leading to transparency because transaction details are available to everyone (Catalini and Gans, 2016). The visibility of the data also assuages the apprehensions of fraud as all the participants get an opportunity to check the authenticity of the data.
3. **Transactions and Blocks:** Intuitively, the actions that occur in a blockchain are organized in blocks that are connected in the order of their occurrence, forming a chain. Two basic consensus models, PoW and PoS, are used to authenticate these transactions before they are integrated into the chain. For instance, PoW involves nodes to validate blocks through a solution of cumbersome mathematical computations as opposed to PoS, where selected validators are determined by their possession of the cryptocurrency and hence are less power-hungry (Nakamoto, 2008).
4. **Cryptography:** Blockchain uses cryptographic hashing to manage transactions, and every block contains a hash of another block. This link ensures that no data within the chain can be easily modified in a way that will affect the authenticity of the whole blockchain (Narayanan et al., 2016).
5. **Consensus Mechanisms:** Every blockchain network has consensus mechanisms that are used to validate transactions taking place on the network. There are several popular consensus algorithms whose primary functionality is to achieve an equal amount of work from all nodes in the network; the best known and most used is PoW, Bitcoin's consensus algorithm. Proof of Stake is another consensus algorithm, where for example, Ethereum uses, and in contrast to PoW, it selects nodes that invest in the network with the help of stakes, saving much energy (Buterin, 2014).
6. **Smart Contracts:** Smart contracts refer to digital contracts that automatically execute themselves when placed in the blockchain. Such contracts include terms that immediately trigger

specific actions for them to be executed without any middleman. For instance, smart contracts can be used to pay for digital delivery of goods, which leads to less conflict and higher operations, as proposed by Szabo (1997). With the help of blockchain's possibilities, smart contracts help complete secure and unchangeable agreements.

12.3.2 THE APPLICATIONS ARE NOT LIMITED TO CRYPTOCURRENCIES

Known as the technology behind digital currencies such as Bitcoin and Ethereum, blockchain is more than just a finance tool. Currently, blockchain is applied to digital identity authentication, voting systems, managing supplies and chains, and healthcare data. A range of these applications is said to realize the upper hand of applying blockchain technology, given that it can impact the improvement of transparency, reduce expenses, and improve traceability (Tapscott and Tapscott, 2016).

12.3.3 TYPES OF BLOCKCHAIN NETWORKS

Blockchain networks are categorized into three primary types: public, private, and consortium blockchains. They are all different services developed for various purposes and have different approaches to governance which includes:

1. **Public Blockchains:** Traditional public channels, such as Bitcoin or Ethereum, are 100% decentralized and include everyone with Internet access in transaction verification or mining. Due to their decentralized nature, these networks are relatively immune to censorship, making them suitable for cryptocurrencies and decentralized applications (dApps) (Nakamoto, 2008).
2. **Private Blockchains:** Private blockchains include Hyperledger Fabric and Corda, which are centralized solutions operated by one organization. Unlike social networks, these networks are centralized, giving more control over transactions while users enjoy more privacy. Companies typically employ a private blockchain to operate on sensitive information, in contrast to public blockchains

which have higher throughput and lower numbers of transactions (Androulaki et al., 2018).

3. **Consortium Blockchains:** In consortium blockchains, many organizations participate as managers of the blockchain network – providing a more decentralized approach for a fairly centralized management. These networks are most relevant for a market that consists of multiple members per node, such as banking, international payments, or energy trading. Such consortium blockchain samples include R3 and the Energy Web Foundation. These networks are most useful for companies that require infrastructure that allows for secure and transparent transactions that can be completed in an efficient manner, making these network choices well suited for industries where collaboration and trust between participants is paramount (Gartner, 2019).

Blockchain technology has significant promising applications in different fields because of its innovative nature – decentralized, secure, and transparent. From Bitcoin to the Ethereum platform, smart contracts, and many others, blockchain is a new thought leader in breaking traditional systems, reducing intermediate parties, making processes faster, and building trust. The category of blockchain can be classified as public, private, and consortium, which adapt flexibility of application from cryptocurrency to enterprise data. This, for every single business, implies the higher levels of innovative disruption that the further evolution of blockchain systems will exert.

12.3.4 HOW BLOCKCHAIN WORKS, INCLUDING CONCEPTS OF DECENTRALIZATION, CONSENSUS MECHANISMS, AND SMART CONTRACTS

Pursuing the goals of enhancing security and trust, blockchain is implemented differently from centralized systems, incorporating key ideas such as decentralization, consensus mechanisms, smart contracts, and more (Nakamoto, 2008). Below is a deeper examination of the relationship between these components.

12.3.4.1 DECENTRALIZATION

1. **Foundation:** Contrary to the concept of a central body that manages the entire process, blockchain operates under distributed computing systems commonly referred to as nodes. In this version, there is no central authority, as every node holds a complete record of the ledger.
2. **Implications:** This structure avoids monopolization of the network by one actor, thus increasing the network security and reducing the risks of coordinated attacks on potential central hubs. It also favors the transparency of a record since every transaction and its checking and approval are open to the network participants.

12.3.4.2 TRANSACTION PROCESSING

1. **Initiation:** It is the action begun by the user where they can send a transaction that includes sending digital currency, performing smart contracts, or even writing data on the blockchain.
2. **Broadcasting:** The transaction is then transmitted over the network, waiting to be validated by the network's nodes.
3. **Validation:** Nodes do this by checking the transaction's contents and the sender's balance (for currency exchanges).

12.3.4.3 CONSENSUS MECHANISMS

1. **Purpose:** These procedures are relevant for maintaining the network's security. They ensure accuracy in the current state of nodes and also the reality of undertaken transactions in the blockchain.
2. **Examples:**
   a. **Proof of Work (PoW):** Bitcoin employs PoW to mitigate fraud as it only allows new transaction blocks to be added if nodes successfully solve mathematical riddles.

b. **Proof of Stake (PoS):** This system chooses independent validators through their stakes on specific cryptocurrency and their allegiance to a certain network, and it requires lesser resources than PoW.

12.3.4.4 BLOCK FORMATION

1. **Creating Blocks:** A verified transaction is bundled into several blocks, of which the hash of the previous block links all of them in a chain.
2. **Adding to the Chain:** These are appended to the blockchain, revise the ledger across the network, and strengthen the records against any alteration.

12.3.4.5 SMART CONTRACTS

Definition: Smart contracts are contractual arrangements that, upon activation, execute operations on the blockchain. Once conditions involving them have been met, they trigger actions without human intervention.

Functionality: For instance, instead of intermediary processing payments and handling disputes between a buyer and a seller, the smart contract can pay the seller upon confirmation that the goods have been delivered.

Benefits: Intelligent contracts built on blockchain offer optimal and safeguarded deals mainly because of smart contracts' unique attributes of security and openness.

12.3.5 OVERALL WORKFLOW

Commencing a transaction, processing the consensus, freezing this transaction in a block, and adding the block to the chain is the basic operational cycle of a blockchain. On the other hand, smart contracts execute the terms of an agreement in an efficient and secure manner, consensus mechanisms validate and safeguard transactions and ensure the network coheres, while decentralization rules out the existence of a central control authority. These characteristics make blockchain a very stable technology that can be applied in many ways in addition to the financial market. Private blockchains are an enterprise version of public blockchains where consortium blockchains are a subset of private blockchains. There are three major forms of blockchain technology that vary primarily by the permissions and management that regulate the member networks: public, private, and consortium. There are four types of audit work, and each is employed in a particular way depending on the type of sector that is involved. Public, private, and consortium blockchains represent different types of blockchain networks, each with unique characteristics and use cases are discussed in the subsections.

12.3.5.1 PUBLIC BLOCKCHAINS

1. **Examples:** Bitcoin, Ethereum.
2. **Characteristics:**
    a. **Decentralization:** The network has full decentralization; nobody owns the network, which increases its security and immunity to censorship.
    b. **Accessibility:** There are no restrictions for anyone to join and engage themselves in some of the procedures, such as transaction confirmation or mining.
3. **Use Cases:** Originally designed for decentralized digital assets such as cryptocurrencies and decentralized applications, also known as dApps.

12.3.5.2 PRIVATE BLOCKCHAINS

1. **Examples:** Hyperledger Fabric, Corda.
2. **Characteristics:**

a. **Centralization:** Coordinated and operated under the administration of one organization rather than several organizations.

b. **Privacy:** Enables transaction processing to be entirely within the organizational system, satisfying business needs where information privacy is vital.

c. **Efficiency:** Usually more efficient than public blockchains because of the proportions of transactions and necessary nodes for consensus.

3. **Use Cases:** These are ideal for companies where they may be used in supply chain management, internal voting, or other areas where privacy is an important factor.

12.3.5.3 CONSORTIUM BLOCKCHAINS

1. **Examples:** R3, the company consisting of over 200 financial institutions, and Energy Web Foundation.

2. **Characteristics:**

a. **Decentralization:** Jointly controlled by a group of institutions, thereby reducing the occurrence of central control problems.

b. **Privacy and Efficiency:** This is a form of compromise between the decentralized public blockchains and the centralized, also more private, but efficient private blockchains.

c. **Controlled Access:** Access and participation are limited to the members of the organization needing the site only, although some sites may allow more public access.

3. **Use Cases:** Perfect for scenarios in which several parties, the banking sector, for instance, or the energy market, must securely exchange information or execute transactions.

These different types of blockchain work according to need and provide various opportunities in different fields, from a fully open and decentralized block chain to restricted and private access.

12.4 BLOCKCHAIN IN FINANCIAL MARKETS

Through the emergence of new transaction methods, reduction of costs, and improvements to the transparency and security of financial markets, blockchain technology has had a significant impact on financial markets around the world. Here are some critical applications of blockchain in financial markets.

12.4.1 ELECTRONIC CURRENCIES (BITCOIN, ETHEREUM, AND OTHERS)

1. **Overview:** Cryptocurrencies are virtual money which employs encryption techniques for security measures and is also based on blockchain. They act as peer-to-peer money in the system.
2. **Impact:** Cryptocurrencies have played a huge part in changing monetary systems, and people now have another choice apart from the traditional fiat currencies. They enable P2P exchange with no need for the intervention of bodies such as banks, which may eliminate fees and increase transaction velocity.
3. **Use Cases:** In addition to being digital monies, cryptocurrencies facilitate dApps on Ethereum's platform for smart contracts and digital tokens representing ownership or claim on an asset.

12.4.2 CROSS-BORDER PAYMENTS

1. **Overview:** Cross-border payments get a boost through blockchain since the process does not require multiple intermediaries like before.
2. **Impact:** Blockchain makes these transactions seamless, much cheaper, and faster. They can complete near real-time and cheaper global transactions than the traditional approaches.
3. **Use Cases:** Blockchain makes RippleNet and other financial entities and services provide payment solutions that are faster, more transparent, and more affordable than other existing solutions.

12.4.3 SECURITIES SETTLEMENT

1. **Overview:** Securities settlement relates to the transfer of securities from the seller to the buyer and the proper payment, which effectively takes several days.
2. **Impact:** Here, blockchain eliminates fragmentation of the transaction record by automating and accelerating the securities settlement process to near real-time, minimizes counterparty risk, and boosts liquidity.
3. **Use Cases:** Many blockchains are built to automate the back office, post-trade, reporting, and clearance of equities, fixed income, and other financial products.

12.4.4 DIGITAL IDENTITY VERIFICATION

1. Impact: Blockchain ensures a secure platform for storing and disclosing individual data and documents based on their choice.
2. Use Cases: Financial services providers and emerging fintech players use blockchain to enhance and protect identity verification procedures, which can enable them to experience improved customer acquisition and curb the probability of identity fraud.

12.4.5 SUPPLY CHAIN FINANCE (SCF)

Overview: It acts as a financing tool for particular products going through the manufacturing to the retail level of the supply chain, thus enhancing the cash flow and minimizing the required capital.

Impact: Supply Chain—Blockchain brings transparency, auditability, and smart execution to the supply chain environment. It helps the stakeholders track the goods and transaction progression so that the risk assessment is more realistic and funds are better utilized.

Use Cases: Blockchain platforms improve SCF by bringing transaction transparency, increased supplier payments, pay-as-you-go, and minimizing fraud. This is especially important for securing transactions and accelerating the disbursement of funds.

12.4.6 BENEFITS OF USING BLOCKCHAIN IN THESE AREAS (E.G., INCREASED TRANSPARENCY, SECURITY, AND EFFICIENCY).

Challenges of effort estimation in such fields (For instance, blockchain application in such areas as (a) Benefits that come alongside the use of blockchain such as increased transparency, security, and efficiency.

Application of the blockchain technology in the financial market area has many advantages for each branch of this market connected with increasing the transparency, security, and effectiveness. Here's how these advantages unfold across different applications.

12.4.6.1 INCREASED TRANSPARENCY

1. **Cryptocurrencies:** Since the ledger is public, each transaction is recorded with all other participants, meaning that users have high trust in the network, and there is little to no fraud.
2. **Cross-border Payments:** First, blockchain ensures the openness of all transactions throughout the payment process through an auditable and transparent transactional trail, which helps track the source of funds while making the clarification of fees easy.
3. **Securities Settlement:** Blockchain allows all the members involved in the trading avenue to have direct SAC access to transaction history and settlements with no question of errors or discrepancies that occur with traditional systems.

12.4.6.2 ENHANCED SECURITY

1. **Cryptocurrencies:** Blockchain is highly secure due to its decentralized system and cryptographic hash function, which helps prevent hacks and fraudulent transactions. Any alteration of the data would mean going against the majority opinion of the entire network, making cryptocurrencies more resistant to alteration.
2. **Digital Identity Verification:** Identity data being stored on the blockchain eliminates certain common vices such as identity thefts

and frauds, for the simple reason that data is not stored centrally and hence cannot be easily prone to a hacker's attack.

3. **Supply Chain Finance:** Because the record entered into the block cannot be manipulated once recorded, blockchain maintains accurate and authentic records for goods and business transactions in the supply chain.

12.4.6.3 INCREASED EFFICIENCY

1. **Cross-border Payments:** Blockchain minimizes cross-border payments by removing the need for middlemen, resulting in shorter processing times from days to minutes and seconds, and lower costs.
2. **Securities Settlement:** Simplification of the trading process through smart contracts, and the sustenance of a permanent record of the dealings also significantly reduces the time required to clear and settle securities trading, hence enhancing liquidity in the market.
3. **Supply Chain Finance:** Realize goods and transactions by implementing real-time tracking, so that there is a chance to speed up invoicing and payment. This helps in early expansion of funds and lowers the business working capital cycle.

Altogether, it means that the application of the technologies based on the blockchain in financial markets allows them not only to be more transparent, secure, and efficient, but also creates the basis for the more effective distribution of financial services. Besides, it stimulates the further evolution of the traditional financial systems. These transformational advantages will fan the flame to a more stable, effective, and diversified financial sector for all players.

12.5 CASE STUDIES

The development of fintech and the usage of blockchain technologies enable rebuilding the financial market priorities as companies see the benefits of applying new tools in solving traditional problems and creating new opportunities. Below are a few case studies that illustrate these trends and applications.

12.5.1 CASE STUDY 1 – RIPPLE AND CROSS-BORDER PAYMENT

1. **Overview:** Cross-border payments is another area where Ripple is applying blockchain in its truest sense through its digital currency XRP, which acts as an intermediate currency to facilitate near real-time, very low-cost transactions.
2. **Application:** RippleNet, the provider of Ripple, is a blockchain system that facilitates peer-to-peer transactions between institutions without help from middlemen, thus reducing expenditures and time likely to be incurred in performing the usual processes.
3. **Impact:** As Ripple connects with over 300 partners of financial services in 40 countries, blockchain is demonstrated to work at scale and deliver an improvement in operational performance as well as customer experience by providing a faster and less costly way of completing transactions.

12.5.2 CASE STUDY 2 – ETHEREUM AND DECENTRALIZED FINANCE (DEFI)

1. **Overview:** Ethereum contracts allow for smart contracts and decentralized applications, or dApps, which are core to fostering the Decentralized Finance or DeFi movement, which envisions a world with no middlemen to traditional financial actors.
2. **Application:** Ethereum-based DeFi platforms offer a range of financial services, including denied lending, borrowing, and trading services, over the internet on a decentralized basis globally.
3. **Impact:** DeFi has emerged as the market has rapidly developed, with billions of dollars within its contracts, demonstrating the ability of blockchain to increase access to financial services, provide more transparency, and significantly reduce counterparty risks.

12.5.3 CASE STUDY 3 – AS PART OF ITS BLOCKCHAIN NARRATIVE, IBM BLOCKCHAIN ALSO PROVIDES SUPPLY CHAIN FINANCE

1. **Overview:** IBM Blockchain proves that blockchain is not only useful when it comes to tangible financial operations, but particularly when it comes to supply chain finance, as the platform provides a transparent and highly secured system to generate an accurate product history.
2. **Application:** From these, IBM solutions offer real time visibility of goods and transaction flows that can effect the supply chain, reduce fraud, and improve payment processes.
3. **Impact:** We assess that cooperation between IBM and Maersk brought the birth of TradeLens, a blockchain shipping solution that has already handled millions of shipments and significantly decreased the time required to process documents related to the supply chain while also cutting overall shipping costs.

12.5.4 CASE STUDY 4 – THIS PAPER LOOKS AT BITPESA AND FINANCIAL INCLUSION

1. **Overview:** BitPesa mainly uses blockchain to offer a somewhat distinct payment service targeting the African region. It will provide a more streamlined solution for users to conduct payments in local and digital money.
2. **Application:** Unlike many other cross-border payment service providers, BitPesa does not involve conventional banking channels, which makes their services cheaper and faster.
3. **Impact:** BitPesa has the potential to show that blockchain can be of unique value in Africa and other regions with low levels of financial inclusion by expanding access to payments services and demonstrating how existing players in this space can use blockchain solutions to improve their operations and access to markets.

The presented case studies illustrate how and where fintech and blockchain have made a difference in the financial market. It both suppresses inefficiencies and costs, and generates new capabilities for financial inclusion and innovation to contribute to the improved efficiency, transparency, and inclusion of the global financial system.

12.6 CHALLENGES AND LIMITATIONS

Although fintech and blockchain technology have come a long way in reinventing the global financial markets, they still have several challenges and constraints that require further exploration for deeper mainstream implementation. Here are some of the main challenges and constraints associated with fintech and blockchain applications in financial markets:

1. **Scalability Issues:**
   a. **Blockchain Scalability:** Current public blockchains such as Bitcoin and Ethereum require assistance with scaling, as they are limited in their transaction throughput due to decentralizing transaction handling, resulting in congestion and high fees during rush hours.
2. **Regulatory Challenges:**
   a. **Regulatory Uncertainty:** The fintech and blockchain industry is relatively young and dynamic, which means that it can quickly progress beyond existing legal standards and leave companies uncertain or restricted on account of the legal grey areas.
   b. **Compliance Costs:** Different jurisdictions' anti-money laundering (AML) and know your customer (KYC) regulations are difficult and expensive for fintech firms to implement.
3. **Security and Privacy Concerns:**
   a. **Security Risks:** However, the ecosystem connected to the technology, such as exchanges and wallets, continues to remain an open target for hackers, eroding trust in these ideas.
   b. **Privacy Issues:** Public blockchains erode privacy due to the openness of the underlying infrastructure, where, while transaction details are anonymous to some extent, they can be directly associated with identity.

4. **The Challenge of Integration with Existing Systems:**
   a. **Interoperability:** Incompatibilities between traditional financial systems and other blockchains are a major drawback of blockchain systems, including their applicability.
   b. **Legacy Systems:** Implementing new fintech solutions into financial institutions' existing frameworks can be difficult, with additional cost and time constraints.
5. **Market Adoption and Trust:**
   a. **User Trust and Adoption:** Gaining users' trust to adopt new fintech and blockchain services is challenging due to security concerns, privacy concerns, and cryptocurrencies' volatility.
   b. **Understanding and Education:** A significant knowledge gap about the benefits and operations of blockchain technology exists among potential users and within the financial sector, which can slow adoption and development.
6. **Environmental Concerns:**
   a. **Energy Consumption:** Some blockchain implementations, particularly those using PoW consensus mechanisms, are highly energy-intensive, leading to environmental concerns about their sustainability.
   b. **Strategies for Overcoming Challenges:** To address these challenges, a cooperative effort among tech developers, regulators, financial institutions, and users is essential. Innovations like more scalable consensus mechanisms (e.g., Proof of Stake), efforts to harmonize regulatory frameworks, and initiatives to enhance interoperability and security are vital for reducing these limitations and promoting the successful integration of fintech and blockchain into financial markets.

12.7 FUTURE TRENDS AND PREDICTION

The convergence of fintech and blockchain in the financial sector is continuously evolving, introducing innovative trends and applications poised to redefine the industry's future. Here are some pivotal trends and forecasts are discussed in the subsections.

12.7.1 WIDER ADOPTION OF BLOCKCHAIN IN CONVENTIONAL FINANCE

1. **Forecast:** Traditional financial entities will increasingly integrate blockchain for various uses, such as cross-border transactions, securities settlements, and regulatory adherence, driven by the demand for enhanced efficiency, security, and cost reduction.
2. **Usage:** Expect to see more financial institutions and services launching or collaborating with blockchain platforms to optimize operations and bolster security.

12.7.2 GROWTH OF CENTRAL BANK DIGITAL CURRENCIES (CBDCS)

1. **Forecast:** The development and eventual deployment of CBDCs by global central banks are anticipated to accelerate, potentially transforming monetary policies and financial frameworks.
2. **Usage:** CBDCs might offer a stable, secure digital currency alternative, boosting payment and settlement efficiencies and possibly expanding financial inclusion.

12.7.2 PROGRESS IN SMART CONTRACTS

1. **Forecast:** The capabilities and intricacies of intelligent contracts are expected to progress, facilitating more complex and specific financial transactions, paving the way for new financial products and services.
2. **Usage:** This could automate financial instruments like loans, insurance, and derivatives, enhancing efficiency and reducing intermediary reliance.

12.7.4 INCREASED TOKENIZATION

1. **Forecast:** Representing tangible assets as blockchain tokens will likely change how assets are bought, sold, and managed, making markets like real estate, art, and collectibles more accessible and liquid.

2. **Usage:** New platforms and exchanges for trading tokenized assets could emerge, widening investment opportunities for investors.

12.7.5 IMPROVED PRIVACY AND SECURITY MEASURES

1. **Forecast:** As blockchain and fintech become more prevalent, focusing on privacy and security enhancements will be crucial, and novel solutions will be developed to safeguard data and transactions.
2. **Usage:** Advancements in encryption, secure multiparty computation, and zero-knowledge proofs could enable secure transaction processing and information sharing on the blockchain without revealing sensitive data.

12.7.6 INTEGRATION OF ARTIFICIAL INTELLIGENCE (AI) WITH BLOCKCHAIN

1. **Forecast:** The synergy between AI and blockchain is expected to grow, enhancing data analysis, decision-making, and automation.
2. **Usage:** AI could improve blockchain efficiency through optimized mining processes, enhanced intelligent contract functionality, and tailored financial services.

12.7.7 RISE OF DECENTRALIZED FINANCE (DEFI)

1. **Forecast:** The DeFi sector, aiming to redesign traditional financial systems with decentralized structures, is projected to expand and potentially integrate more with conventional finance.
2. **Usage:** DeFi platforms might provide alternatives to nearly all existing financial services, such as savings, loans, trading, and insurance, without traditional financial intermediaries.

12.7.8 REGULATORY EVOLUTION AND STANDARDIZATION

1. **Forecast:** As fintech and blockchain mature, regulatory frameworks are likely to adapt to better manage and standardize these innovations across regions.

2. **Usage:** More precise regulations could foster broader blockchain adoption by offering legal clarity and protections for financial market users and providers.

These trends underscore the transformative capacity of fintech and blockchain in redefining financial markets, offering more secure, efficient, and accessible services, though they also pose regulatory, technical, and adoption challenges.

12.8 CONCLUSION

The convergence of fintech and blockchain in the financial sector represents a pivotal shift, not merely a temporary trend, fundamentally transforming financial services. This integration is manifesting through various applications, from cryptocurrencies to cross-border payments, securities settlements, digital identity verifications, and supply chain financing, underscoring blockchain's broad potential.

12.8.1 KEY TRENDS

1. **Decentralization of Financial Services:** At its core, blockchain promotes decentralization, disrupting traditional centralized financial models and advocating a more equitable system where control is dispersed throughout the network.
2. **Enhanced Security and Transparency:** Blockchain enhances the security and transparency of transactions, crucial in an era dominated by concerns over data breaches and financial fraud. Its immutable ledger and cryptographic safeguards offer solid defenses against these threats.
3. **Operational Efficiency and Cost Reduction:** By streamlining processes and removing intermediaries, blockchain significantly reduces costs and boosts operational efficiency, particularly noticeable in cross-border transactions and securities settlements.
4. **Innovation in Financial Products and Services:** Blockchain's adoption is spurring new financial products and services. From

cryptocurrencies redefining money to smart contracts automating and enforcing terms, blockchain fosters financial innovation.

5. **Regulatory Evolution:** As blockchain gains traction, global regulatory frameworks are adapting to support and regulate this technology, a critical step for its mainstream integration and safe, ethical utilization.

Applications and Impact Blockchain, fueled by fintech innovations, not only increases transactional efficiency, security, and transparency but also promotes financial inclusion, offering essential services to the unbanked and underbanked. Its roles in digital identity verification and supply chain finance demonstrate its capability to transcend traditional financial boundaries, enhancing security and efficiency across various sectors.

12.8.2 CONCLUSION

The combination of fintech and blockchain is not just hype, but it is one of the industry's more fundamental changes regarding global financial systems. At its core, blockchain has the potential to provide a fairer and less centralized method of delivering financial services, which relies less on mainstream third parties. This shift is advantageous for security because data in a blockchain ledger is not concentrated in a single point and because, once recorded, the data cannot be altered (Tapscott and Tapscott, 2016). Additionally, through a disintermediation of the central authorities as well as deconstruction and integration of various intermediary services, blockchain contributes to the processes and cost optimization, particularly in the areas of international payments and securities trading (Catalini and Gans, 2016).

In addition, through the help of blockchain, financial instruments and services define innovative trends. Blockchain technology is well on the way to shifting the manner in which financial transactions occur through the use of cryptocurrencies, intelligent contracts, and so forth (Nakamoto, 2008). The legal environment is also shifting since international standards are being steadily calibrated to mediate and govern fintech and blockchain tools as these are embraced more broadly (Gomber et al., 2017). Therefore, both fintech and blockchain contribute not only to the amelioration of the efficacy and security of financial sectors but also to reducing Financial

Exclusion by expanding access to financial services to various populations. Due to their prospects for revolutionizing the current financial paradigms, introducing novel solutions, and solving historical issues, they are considered as main actors of future development and change in the financial markets (Puschmann, 2017).

KEYWORDS

- **digital money transfer**
- **digital sale**
- **distributed ledger**
- **financial services**
- **financial technology**
- **stock exchanges**
- **technology**

REFERENCES

Arner, D. W., Barberis, J. N., & Buckley, R. P. (2016). The evolution of Fintech: A new post-crisis paradigm? *Georgetown Journal of International Law*, *47*(4), 1271–1319.

Au, C., Au, T., & Fung, W. (2018). Blockchain-SimGame: How can simulation be used for facilitating corporate learning of blockchain technology? *European Conference on Games Based Learning*, 847–851.

Catalini, C., & Gans, J. S. (2016). Some simple economics of the blockchain. *National Bureau of Economic Research*. https://doi.org/10.3386/w22952 (accessed on 26th July 2025).

Demi, S., & Sánchez-Gordón, M. (2021). Software engineering applications enabled by blockchain technology: A systematic mapping study. *Applied Sciences*, *11*(7), 2960. https://doi.org/10.3390/app11072960 (accessed on 26th July 2025).

Fama, E. F. (1970). Efficient capital markets: A review of theory and empirical work. *Journal of Finance*, *25*(2), 383–417.

Gomber, P., Koch, J. A., & Siering, M. (2017). Digital finance and Fintech: Current research and future research directions. *Journal of Business Economics*, *87*(5), 537–580. https://doi.org/10.1007/s11573-017-0852-x (accessed on 26th July 2025).

Goetzmann, W. N. (2005). *The Origins of Financial Markets*. Oxford University Press.

Haddad, C., & Hornuf, L. (2019). The emergence of the global Fintech market: Economic and technological determinants. *Small Business Economics*, *53*(1), 81–105. https://doi.org/10.1007/s11187-018-9991-x (accessed on 26th July 2025).

InsurTech synergy: How software is revolutionizing insurance processes–Medicalrchi Tecture. (n.d.). *Medicalrchitecture*. https://medicalrchitecture.com/insurtech-synergy-how-software-is-revolutionizing-insurance-processes (accessed on 26th July 2025).

The rise of Luna crypto in the market. (n.d.). *CryptoTradeSignals.live*. https://cryptotradesignals.live/article/article.php?article=3-the-rise-of-luna-crypto-in-the-market&id=207053 (accessed on 26th July 2025).

Khan, A. (2021). BlockPres: A novel blockchain-based incentive mechanism to mitigate inequalities for prescription management system. *Sensors*, *21*(15), 5035. https://doi.org/10.3390/s21155035 (accessed on 26th July 2025).

Lee, I., & Shin, Y. J. (2018). Fintech: Ecosystem, business models, investment decisions, and challenges. *Business Horizons*, *61*(1), 35–46. https://doi.org/10.1016/j.bushor.2017.09.003 (accessed on 26th July 2025).

Madhavan, A. (2000). Market microstructure: A survey. *Journal of Financial Markets*, *3*(3), 205–258. https://doi.org/10.1016/S1386-4181(00)00007-0 (accessed on 26th July 2025).

Markowitz, H. (1952). Portfolio selection. *The Journal of Finance*, *7*(1), 77–91. https://doi.org/10.2307/2975974 (accessed on 26th July 2025).

Nakamoto, S. (2008). *Bitcoin: A Peer-To-Peer Electronic Cash System*. Bitcoin.org. https://bitcoin.org/bitcoin.pdf (accessed on 26th July 2025).

Neal, L. (2005). *The Rise of Financial Capitalism: International Capital Markets in The Age of Reason*. Cambridge University Press.

Pilkington, M. (2016). Blockchain technology: Principles and applications. In *Research Handbook on Digital Transformations* (pp. 225–253). Edward Elgar Publishing.

Puschmann, T. (2017). Fintech. *Business & Information Systems Engineering*, *59*(1), 69–76. https://doi.org/10.1007/s12599-017-0464-6 (accessed on 26th July 2025).

Rajput, N., Garg, V., & Jyotsna. (2023). Fintech & blockchain technology in financial markets: Trends and applications. Sri Aurobindo College, University of Delhi; Amity University Uttar Pradesh, Greater Noida Campus; Jagan Institute of Management Studies.

Schueffel, P. (2016). Taming the beast: A scientific definition of Fintech. *Journal of Innovation Management*, *4*(4), 32–54. https://doi.org/10.24840/2183-0606_004.004_0004 (accessed on 26th July 2025).

Sharpe, W. F. (1964). Capital asset prices: A theory of market equilibrium under conditions of risk. *The Journal of Finance*, *19*(3), 425–442. https://doi.org/10.2307/2977928 (accessed on 26th July 2025).

Tapscott, D., & Tapscott, A. (2016). *Blockchain Revolution: How the Technology Behind Bitcoin is Changing Money, Business, and the World*. Penguin.

Zavolokina, L., Dolata, M., & Schwabe, G. (2017). Fintech transformation: How IT-enabled innovations shape the financial sector. *Electronic Markets*, *27*(3), 231–243. https://doi.org/10.1007/s12525-017-0249-8 (accessed on 26th July 2025).

Index

C

D

G

Q

R

S

T

For Product Safety Concerns and Information please contact our EU
representative GPSR@taylorandfrancis.com
Taylor & Francis Verlag GmbH, Kaufingerstraße 24, 80331 München, Germany

www.ingramcontent.com/pod-product-compliance
Ingram Content Group UK Ltd.
Pitfield, Milton Keynes, MK11 3LW, UK
UKHW021501240926
475702UK00035B/100
9781779641281